The Scopes Trial

The Scopes Trial

An Encyclopedic History

RANDY MOORE *AND*
SUSAN E. BROOKS

McFarland & Company, Inc., Publishers
Jefferson, North Carolina

ISBN (print) 978-1-4766-8544-1
ISBN (ebook) 978-1-4766-4819-4

LIBRARY OF CONGRESS AND BRITISH LIBRARY
CATALOGUING DATA ARE AVAILABLE

Library of Congress Control Number 2022049940

Front cover: Defense attorney Clarence Darrow addresses the jury during the Scopes Trial in July 1925 in Dayton, Tennessee (Tennessee State Library and Archives)

Printed in the United States of America

McFarland & Company, Inc., Publishers
Box 611, Jefferson, North Carolina 28640
www.mcfarlandpub.com

To Tom, Pat, Dean, Ann, Sue, Brenda, Todd, Nigel, Ralph, Freddie, Felicia, Curtis, Jacob, and my other many friends in Dayton, whose help and hospitality for the past 25 years have enriched my life and made a project like this possible.—R.M.

To my mother, Jeanette Gilliam Travis, for instilling a love of old books and family history; my father, Kelsie Don Travis, for encouraging me to always ask questions and try new things; and my husband, Christopher Brooks, our four sons, two daughters-in-law, and five grandchildren. I love you all.—S.B.

Table of Contents

Acknowledgments viii
Chronology ix
Preface 1
Introduction: What Happened in Dayton? 3

The Encyclopedia 9

Appendix: The Scopes Family 197
Notes 199
Bibliography 202
Index 203

Acknowledgments

We have had much help during the past 25 years from our many friends in Dayton, including Pat Guffey, Tom Davis, Dean Wilson, Ann Gabbert Bates, Kevin Woodruff, Ralph Green, Curtis Lipps, Freddie Brewer, Brenda Marler, Sue Purser, Todd Wood, the late Jimmy McKenzie, Donna and Tom Taylor, Nigel Chadwick, Jacob Ellis, the late Eloise Reed, and Lou King. These friends also offered much knowledge about the people, locations, and culture of the area. They, along with other members of the Rhea County Historical and Genealogical Society, also offered camaraderie and new contexts for our discussions and explorations of even the most tangential aspects of the trial.

From elsewhere, we were helped by Ryan Greenwood, Todd Hatton, Susan and Jon Epperson, Jerry Adelman, Steve Smartt, the late Layton Brueske, Vickie Bryant, and Linda Wynn. Judith Ridner, Matthew Lavine, and Alan Marcus provided insights and asked questions that often shifted our way of thinking. Additionally, the librarians and museum curators in Salem, Illinois; Washington, D.C.; Ft. Worth, Texas; Knoxville, Tennessee; Paducah, Kentucky; Bowling Green, Kentucky; and Starkville, Mississippi, were an invaluable help and deserve more praise than we can give here.

We are also especially grateful to John Scopes' friends, children, and other relatives who graciously met with us and hosted us in their homes, including John Scopes, Jr., the late Jackie and Bill Scopes, Lisa Rennegarbe, Nancy Rose, Jerry Tompkins, Jim and Fran Presley, Richard and Patricia Heflin, Barbara Heflin Kough, Nathan Lynn, Nancy Heflin Schur, Jeanette Gilliam Travis, and Regenia Buchanan. Special thanks go to our spouses/partners, friends, students, and families who endured our absences or traveled with us while we conducted research. We also thank Regenia Buchanan, Christopher Brooks, and Andrea Crenshaw for their excellent chauffeuring skills in Paducah and to and from the Scopes Festival. We especially thank Richard and Pat Heflin, Jeanette Gilliam Travis, and the late Bill and Jackie Scopes for access to and the use of their family photographs, unpublished manuscripts, and correspondence. Finally, we thank the many other people who gave us directions and other courtesies whom we cannot remember, but nevertheless contributed to our work. They deserve more praise than our poor memories enable us to give.

Chronology

1859	November 24	In *On the Origin of Species*, Charles Darwin presents his theory of evolution by natural selection. Darwin claims "the view which most naturalists entertain, and which I formerly entertained—namely, that each species has been independently created—is erroneous."
1871	February 24	In *The Descent of Man*, Darwin makes explicit that there was a recent common ancestor between humans and apes.
1886	February 9	Dayton Coal and Iron Company (DCIC) opens its first furnace just outside of Dayton, Tennessee.
1896	July 9	William Jennings Bryan's electrifying "Cross of Gold" speech at the Democratic National Convention in Chicago catapults him to national prominence and the party's nomination for president.
1913		DCIC fails, depressing the economy of Dayton and the surrounding area.
1914		George Hunter, a teacher at DeWitt Clinton High School in New York City, publishes *A Civic Biology: Presented in Problems.* Hunter's high school textbook, which promoted evolution, racism, and eugenics, was adopted in Dayton, used by substitute teacher John Scopes, and cited in Scopes' trial.
1920	July 1	In the Baptist newspaper *The Watchman-Examiner*, Curtis Lee Laws tells readers that people "who cling to the great fundamentals and who mean to do battle royal for the fundamentals shall be called 'Fundamentalists.'"
1924	August	John Scopes arrives in Dayton to coach and teach at Rhea Central High School.
1925	January 21	Tennessee representative John Butler introduces House Bill 185 making it a crime to teach "any theory that denies the story of the Divine Creation of man as taught in the Bible, and to teach instead that man has descended from a lower order of animals."
1925	January 28	Tennessee's House of Representatives passes the Butler Act on a vote of 71 to 5.
1925	March 13	Tennessee's Senate passes the Butler Act on a vote of 24 to 6.
1925	March 21	Tennessee governor Austin Peay signs the Butler Act into law. This is the first law in the United States to criminalize the teaching of human evolution in public schools.

1925	April	Rhea Central High School biologist and principal William Ferguson gets sick and misses two weeks of school. Ferguson's substitute teacher is first-year teacher John Scopes.
1925	May 1	John Scopes' contract ends at Rhea Central High School, but he stays in town to pursue a girlfriend and help two students injured in a car accident.
1925	May 4	Several newspapers in Tennessee report that the ACLU is "looking for a Tennessee teacher who is willing to accept our services in testing [the Butler Act] in the courts. Our lawyers think a friendly test case can be arranged without costing a teacher his or her job.... All we need now is a willing client." George Rappleyea, manager of the remnants of DCIC, reads the article and decides to discuss it with Dayton's leaders.
1925	May 5	Several of Dayton's civic leaders meet in Robinson's Drug Store and begin planning a show-trial to test the Butler Act that they hope will stimulate the area's economy. The group summons John Scopes from a nearby tennis match. Scopes agrees to be the defendant, even though he isn't sure that he taught evolution when he was the substitute teacher in Ferguson's biology class earlier in the year.
1925	May 12	Fundamentalist leader William Bell Riley asks Bryan to represent his World's Christian Fundamentals Association at John Scopes' upcoming trial. Bryan accepts the invitation, thereby ensuring that the trial will attract national media coverage.
1925	May 14	Clarence Darrow, America's most famous criminal defense attorney, offers to help defend Scopes. The ACLU objects, but Scopes insists on being defended by Darrow. This was the only time that Darrow volunteered his services for free to a client.
1925	May 25	Scopes is indicted by a grand jury (whose term had expired) and his trial is set for July 10. In Chicago, WGN radio prepares to broadcast the trial live, the first such broadcast in American history.
1925	July 6	A federal judge rejects an injunction requested by Scopes' defense team to move Scopes' upcoming trial to a federal court.
1925	July 7	Bryan arrives in Dayton. That evening, at a banquet held in his honor at Hotel Aqua, Bryan describes the upcoming trial as a "duel to the death" between evolution and Christianity, adding, "If evolution wins, Christianity goes."
1925	July 8	Journalist H.L. Mencken arrives in Dayton, where he begins writing a series of reports that are now lauded as journalistic landmarks. Mencken is credited with creating the "Monkey Trial" name for Scopes' trial.
1925	July 9	Darrow arrives in Dayton and is honored at a banquet at Hotel Aqua.
1925	July 10	*State of Tennessee v. John Thomas Scopes* opens in the Rhea County Courthouse in Dayton, Tennessee. Because Scopes had been indicted on May 25 by a grand jury whose term had expired, Judge John Raulston immediately convened another grand jury to indict Scopes again.

The Scopes Trial was held July 10–21, 1925, in the Rhea County Courthouse in Dayton, Tennessee (Randy Moore).

1925	July 12	Bryan tells a crowd of 2,000 people on the courthouse lawn in Dayton about the evils of evolution and that he will oppose Darrow's use of expert witnesses in the trial.
1925	July 14	Darrow objects to the trial's daily prayers, but Judge Raulston overrules the objection.
1925	July 16	Malone delivers his "There Is Never a Duel with the Truth" speech, which Bryan describes as "the greatest speech I have ever heard."
1925	July 17	Judge Raulston bars expert testimony by scientists because it would "shed no light" on whether Scopes violated the Butler Act. Darrow is found in contempt of court. Most reporters, believing that the trial is effectively over, leave town.
1925	July 20	After Judge Raulston moves proceedings to a dais on the courthouse lawn, the defense calls Bryan as an expert witness about the Bible. Darrow questions Bryan for 90 minutes about "fool ideas that no intelligent Christian on earth believes." In response, Bryan condemns Darrow for making a "slur at the Bible."
1925	July 21	Judge Raulston expunges Bryan's testimony from the record and Scopes is convicted. Raulston fines Scopes $100, which Bryan and the ACLU offer to pay.
1925	July 23	Scopes rejects an offer to continue work at Rhea Central High School and, with $2,550 raised by journalists, scientists, and others, soon enrolls in graduate school to study geology at the University of Chicago. He excels in his classes, but drops out less than two years later.
1925	July 26	Bryan dies during an afternoon nap in Dayton.
1925	July 31	Bryan is buried in Arlington National Cemetery beneath the epitaph "He Kept the Faith."
1927	January 15	The Tennessee Supreme Court overturns John Scopes' conviction on a technicality (Scopes' $100 fine should have been set by the jury, not Judge Raulston), but upholds the constitutionality of the Butler Act. This ends Darrow's hoped-for challenge of the constitutionality of Tennessee's Butler Act.
1930	September 18	William Jennings Bryan University (WJBU) opens in the abandoned high school where John Scopes allegedly taught evolution.
1932		The ACLU abandons its seven-year search for volunteers to challenge laws in Arkansas, Tennessee, and Mississippi banning the teaching of human evolution in public schools. The various antievolution laws remained unchallenged until 1965.
1935	October	WJBU offers its first classes at its current hilltop location.
1955	January 10	*Inherit the Wind* premiers to favorable reviews at Theatre '55 in Dallas, Texas. The play, which was meant to be a commentary on political intolerance of the then-contemporary McCarthy era, portrays Drummond (i.e., Darrow) as a champion of free speech, and the residents of fictional Hillsboro (i.e., Dayton) as religious zealots. Few people note the disclaimer in the preface: "*Inherit the Wind* is not history."

1955	April 21	*Inherit the Wind* opens on Broadway at National Theater.
1960	July 21	John Scopes returns to Dayton for Scopes Trial Day, where festivities conclude with the North American premier of the movie *Inherit the Wind* at the Dayton Drive-In Theater.
1967	May 18	Tennessee governor Buford Ellington signs legislation repealing the Butler Act.
1967	June	John Scopes and James Presley publish *Center of the Storm: Memoirs of John T. Scopes*, which becomes a regional bestseller.
1968	November 12	In *Epperson v. Arkansas*, the U.S. Supreme Court strikes down an Arkansas law banning the teaching of human evolution in public schools.
1970	April 1	Scopes makes his first appearance in a Tennessee classroom in 45 years when he speaks at Peabody College in Nashville. This was Scopes' last public appearance.
1970	October 21	John Scopes dies of cancer in his home in Shreveport, Louisiana. Scopes' last words were to his wife Mildred: "I love you."
1973	April 30	Tennessee passes its "Genesis Law," which requires public schools to equally emphasize "the Genesis account in the Bible" with other theories about human origins.
1975	August 20	Tennessee's "Genesis Law" is declared unconstitutional.
1976		Rhea County Courthouse, built in 1891, is made a National Historic Landmark.
1978	October	The Scopes Trial Museum opens in the basement of the Rhea County Courthouse.
1988	July	Dayton hosts the first Scopes Trial Play and Festival. These celebrations, which include a reenactment of the trial, have been repeated almost every year since.
2005		Bryan College gives Rhea County a statue of William Jennings Bryan, which is placed on the Rhea County Courthouse lawn.
2017	July 14	A statue of Clarence Darrow, funded with private donations (e.g., from the Freedom from Religion Foundation), is unveiled on the Rhea County Courthouse lawn near the statue of Bryan.
2021	March 1	Court is held for the last time in Rhea County Courthouse. All proceedings were then permanently moved to the Rhea County Justice Center.
2025	July	Dayton celebrates the 100th anniversary of the Scopes Trial.

Preface

Why another book about the Scopes Trial?

There are several good books that narrate the Scopes Trial, and we have used some of these books when writing this book. However, this encyclopedia is not a mere retelling of Scopes' long-studied "Trial of the Century." Indeed, we have added much new information about John Scopes' famous trial that we've gathered from libraries, employers, schools, and forgotten scrapbooks, family files, and newspaper articles in the attics of John Scopes' family, relatives, and friends. In 2015, Scopes Trial historian Todd Hatton (2015) described Scopes as "the one player in the Monkey Trial that people know the least about," and we hope our book helps fill that gap in our understanding of a trial that the *Chattanooga Daily Times* described as being unlike any case "in the annals of Tennessee jurisprudence, or of the world, for that matter" (Small, 1925).

This book—the first encyclopedic treatment of the Scopes Trial—presents the Scopes Trial in a concise, objective, accessible, and new way. It (1) can be read selectively, one entry at a time and in any order, and (2) is not a start-to-finish narrative, but instead consists of integrated descriptions of the people, places, events, and ideas that produced (and were produced by) the trial. Throughout the book, you'll get glimpses of the personal lives and social contexts of the players in the trial—how, when, and where they lived, how they were involved in and/or affected by the trial, and what became of them after the trial. Some of the trial's players became famous, others became martyrs, and still others disappeared from public life. Some became rich, while others endured poverty, prison, murder-for-hire plots, and even suicide. A few are honored by statues, parks, schools, and other monuments, while others rest in unmarked, forgotten graves.

Several excellent biographies already exist about Bryan and Darrow, but relatively little has been published about other important players such as Dudley Field Malone, Mary Baird Bryan, George Rappleyea, Howard Byrd, A.P. Haggard, and Walter White. We hope our book remedies some of these oversights. You'll learn new things in fresh contexts about these and other people. Finally, our book's exhaustive index will help you appreciate the many connections of the people, places, and ideas involved in the trial.

Sources

Hatton, Todd. 2015. I Am John Scopes. Retrieved November 1, 2019, at https://www.youtube.com/watch?v=x9bhXVP06lk; see also www.wkms.org/term/scopes#stream/o.

Small, Robert T. 1925. Scopes juryman kept in dark. *Chattanooga Daily Times*, July 21, 1–2.

Introduction

What Happened in Dayton?

A century ago, America was changing. "The Roaring Twenties" were in full swing. Radios were transforming communication and entertainment, and the first cars coming off assembly lines were freeing people to travel like never before. People were abandoning traditional rural lifestyles for urban manufacturing centers, and a new kind of music called jazz was luring people into nightclubs. Women's dresses were getting shorter, the Ku Klux Klan had been reborn, and troops from the Great War (later called World War I) were coming home with new perspectives on life. Throughout the country, the growing popularity of telephones enabled people to discuss these societal changes with other people, both near and far.

Religious institutions were also changing. Many urban churches and seminaries began embracing modernism, a liberal theology in which the arbiter of truth was logic, not Biblical literalism. According to modernists, miracles such as the virgin birth and Noah's flood had not actually happened, but instead were merely pre-scientific explanations of events that people did not understand. Modernists proudly relied not on literal interpretations of the Bible, but on reason; instead of a six-day "Creation Week," modernists endorsed Darwin's ideas about evolution. To many Christians, modernists had—according to one of John Scopes' prosecutors—made the Bible "a scrap of paper" that was to be accepted, rejected, or amended "according to the whim of the reader" (Bryan, 1924). For traditional Christians, this was heresy.

In the wake of modernism came a religious backlash called "fundamentalism" that emphasized populism and traditional Christian interpretations of the Bible. Although the malleability of theological modernism made it an elusive enemy, its endorsement of evolution provided a concrete, simple target that fundamentalists could easily condemn and attack. Fundamentalists, who were proudly intolerant of modernism, were led by fiery, popular preachers such as William Bell Riley and Billy Sunday, who became celebrities as they warned America of the anarchy, societal corruption, and impending doom that would accompany the teaching of evolution. According to fundamentalists, the teaching of evolution was responsible for *all* of America's problems.

In March 1925, legislators in Tennessee responded to fundamentalists' concerns by passing a law criminalizing the teaching of human evolution in the state's public schools. Tennessee's legislators hoped that their newly passed law would eliminate evolution's threat to morality and religious faith in the same way that in 1920 the societal threats of alcohol were purportedly eliminated by Prohibition.

Amidst these societal upheavals came the Scopes Trial, a landmark event in American history that "enlisted the attention of high brows and low brows the country over" (Anderson, 1925).

The Trial of John Scopes

In early May 1925, John Thomas Scopes—a 24-year-old coach and teacher in Dayton, Tennessee—was charged with violating the state's newly passed anti-evolution statute. Scopes was defended by agnostic Clarence Darrow, America's most revered and reviled criminal lawyer. Darrow, who had previously defended murders, Communists, and other unpopular clients, championed free thought while he ridiculed fundamentalists' attempts to use a state law to ban the teaching of a natural law.

Helping Tennessee prosecute Scopes was the fervently religious, three-time presidential candidate William Jennings Bryan, who promised his followers that John Scopes' trial would be a "duel to the death" between evolution and religious faith. Other fundamentalist leaders issued similar proclamations as they denounced the teaching of evolution as Satanic, destructive, and "utterly incompatible with the Christian religion" (Straton, 1925). Tennessee legislator John Washington Butler, who had authored the law used to prosecute Scopes, warned that the teaching of evolution was "the greatest menace to civilization in the world today," while Texas firebrand preacher Frank Norris, not to be outdone, promised that Scopes' misdemeanor trial would be "the greatest battle of the centuries" (Thompson, 1981). An Oxford University professor even proclaimed that, in his "considered judgment," the Scopes Trial was "the most serious setback to civilization in all history" (Bailey, 1953). Hordes of "screwballs, con men, and devotees ... freaks ... [and] pulpit heroes" descended on Dayton to witness the spectacle (Ballyhooed hullabaloo, 1925). They were not disappointed; when the defense called Bryan to the witness stand on the courthouse lawn, the throng witnessed "the most amazing judicial performance ever witnessed in America"

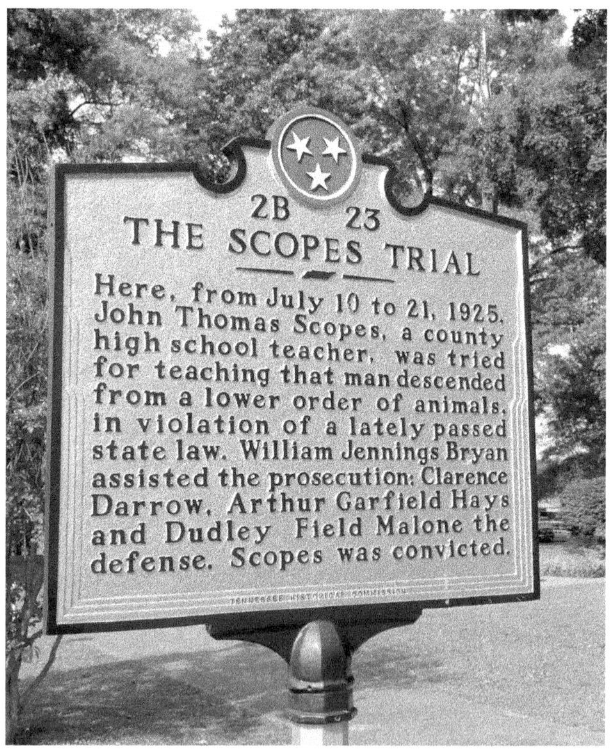

The Scopes Trial is commemorated with this historical marker in front of Rhea County Courthouse. Various error-filled markers had stood here for decades before they were replaced by this accurate marker (Randy Moore).

(Hyde, 1925). According to *Variety*, a magazine devoted to the entertainment industry, John Scopes had become "the biggest current 'name' for any kind of showbusiness" (Offers for Scopes, 1925).

Scopes' innocence or guilt quickly became irrelevant as debates shifted to the relevance of scientists as expert witnesses, the control of the curriculum in public schools, and the tensions between science and religion. As people across America chose sides, Bryan was often portrayed as God's messenger, while Darrow was likened to Satan as they faced off in a courtroom showdown that would decide the fates of Christianity and civilization. A few commentators, realizing that the trial was started for economic reasons rather than legal concerns, were more practical, admitting that the trial in tiny Dayton was "not a fight for evolution or against evolution, but a fight against obscurity" (Doubts legality of special term, 1925). The trial—in which two of the era's best-known personalities clashed over one of society's most controversial subjects—became a world-watched event.

What Happened in Dayton

Since 1925, scientists, politicians, lawyers, and others have seldom tired of writing about the Scopes Trial. There are several good narratives of the trial, so only a brief sketch is presented here for context:

Day 1: Jury Selection
State of Tennessee v. John Thomas Scopes opened on Friday, July 10, 1925, in the Rhea County Courthouse in Dayton, Tennessee. Because Scopes had originally been indicted (on May 25) by a grand jury whose term had expired, Judge John Raulston convened a new grand jury, which deliberated an hour and re-indicted Scopes. The jury was then chosen for the trial.

Day 2: Constitutionality of the Butler Act
Hoping to have the Butler Act declared unconstitutional, Darrow gave a fiery speech claiming that the law violated freedom of religion, describing it as "as brazen and as bold an attempt to destroy learning as was ever made in the Middle Ages." Darrow's speech consumed 13 of the 42 pages of this day's transcript.

Day 3: Debate About Prayer
Darrow objected to the court's opening prayer and the jury being present to hear the discussion of his objection. Judge Raulston overruled Darrow's objection, but instructed preachers offering opening prayers to "make no reference to the issues involved in the case." Raulston then adjourned court to prepare his response to the defense team's claim about the constitutionality of the Butler Act.

Day 4: The Prosecution Presents Its Case
Judge Raulston overruled the defense team's motion to have the Butler Act declared unconstitutional. Raulston ruled that the law "gives no preference to any particular religion or mode of worship," adding, "It is not seriously insisted by the defendant ... that the indictment should be quashed" by the defense's claim that the Butler Act violated the legislature's duty "to cherish literature and science" (from the Tennessee Constitution, Section 12, Article XI). In the afternoon session, Scopes pleaded not guilty, after which Scopes' defender Dudley Field Malone claimed, "While the defense thinks there is a conflict between evolution and the Old Testament, we believe there is no conflict between evolution and Christianity." The prosecution then called its witnesses: Superintendent of Schools Walter White, school board chairperson

F.E. Robinson, student Howard Morgan, and student Harry "Bud" Shelton. The defense team then called its first witness, Johns Hopkins University zoologist Maynard Metcalf.

Day 5: Debate About Expert Testimony

Attorney General Tom Stewart moved to exclude the testimonies of scientists because they would be irrelevant to the case. When the prosecution argued that Stewart's motion was important, defense attorney Arthur Garfield Hays countered that the experts were necessary to teach the jury what evolution is and how it could relate to the Biblical account of creation. Darrow wondered how a jury, only one member of which had ever read about evolution, could intelligently decide the case without hearing from experts. The elder Bryan then denounced the textbook used by Scopes in his biology class and invoked the infamous Leopold and Loeb murder case of 1924 to show the dangers associated with the teaching of evolution. Malone's emotional response to Bryan about truth drew universal praise, including from Bryan. Judge Raulston admitted that the admissibility of the scientific evidence was "the most difficult thing the court is going to pass on."

Day 6: Judge Raulston Rejects the Experts' Testimonies

Judge Raulston excluded the defense team's experts because they could "shed no light" on whether Scopes violated the Butler Act. However, Raulston allowed their statements to be included in the court record for possible appeal. The frustrated Darrow then claimed that the defense had lost "every major point of contention" and implied that Raulston was biased.

Day 7: Darrow Questions Bryan

Judge Raulston cited Darrow for contempt of court, and Hays finished reading the defense team's experts' statements into the court record. Darrow then apologized to the court and was forgiven by Raulston. Claiming that the floor of the second-floor courtroom was faulty, Raulston then adjourned the afternoon's proceedings to an outdoor dais, where a large crowd saw Darrow question Bryan for 90 minutes about the Bible. Raulston then adjourned court for the day. Darrow's questioning of Bryan remains the most famous part of Scopes' trial, as well as a symbol of the ongoing conflicts of religion and science in American society.

Day 8: The Verdict and Sentencing

Rain moved proceedings back inside, where Judge Raulston expunged Bryan's testimony on the previous day from the court record because it could "shed no light upon any issues that will be pending before the higher courts." Darrow then asked the court "to bring in the jury and instruct the jury to find the defendant guilty." Darrow's move, which avoided closing statements, prepared the case for an appeal and saved Darrow, Hays, and Malone from being questioned in court by Bryan. The jury then deliberated nine minutes and found Scopes guilty, after which Raulston fined Scopes $100. Scopes delivered his only statement of the trial, claiming that the Butler Act and his fine were "unjust" and that he would continue "to oppose this law in any way I can. Any other action would be in violation of my ideal of academic freedom—that is, to teach the truth as guaranteed in our constitution, of personal and religious freedom." Neal then motioned for a new trial, which Raulston rejected. The Scopes Trial was over.

Scopes' trial, which one reporter described as "something not of this world," remains known not for its decision, but for the trial itself. Decades after the trial, the largely fictitious movie *Inherit the Wind* added to the Scopes Trial legend and, in the process, provided most people's "knowledge" of Scopes' famous trial.

Although the Scopes Trial symbolized more than it accomplished, it nevertheless became one of the twentieth century's defining events (Puente, 1999). Today, it remains the standard to which all of the many other "evolution trials" have been compared; no other misdemeanor trial in American history has achieved such legendary status. America first became transfixed by a variety of important issues at the Scopes Trial, and we've remained transfixed by them ever since. In this book, you'll learn why.

Sources

Anderson, Paul Y. 1925. The drugstore argument that swelled into a national debate. *St. Louis Post-Dispatch*, May 31, 17.
Bailey, Kenneth K. 1953. The Antievolution Crusade of the Nineteen-Twenties. Ph.D. Dissertation, Vanderbilt University, Nashville, Tennessee, 150.
Ballyhooed hullabaloo. 1925. *Variety*, July 15, 7.
Bryan, William Jennings. 1924. *Shall Christianity Remain Christian? Seven Questions in Dispute*. New York: Fleming H. Revell, 150–151.
Doubts legality of special term. 1925. *Chattanooga Times*, May 24, 9.
Hyde, Henry, M. 1925. Bryan on stand, upholds Bible statements as true. *Baltimore Sun*, July 21, 1.
Offers for Scopes. 1925. *Variety*, July 22, 1.
Puente, M. 1999. The century's biggest news: List ranks headlines of history. *USA Today*, February 25, D1–2.
Straton, John Roach. 1925. *The Famous New York Fundamentalist-Modernist Debates: The Orthodox Side*. New York: George H. Doran, 57.
Thompson, James T. 1981. *Tried by Fire: Southern Baptists and the Religious Controversies of the 1920s*. Macon, GA: Mercer University Press, 131–132.

The Encyclopedia

Alleged (movie)

In 2010, Slingshot Pictures—the "faith-based" division of Imagine Entertainment—released the movie *Alleged*, a "stirring romantic drama set against the landmark Scopes 'Monkey Trial' of 1925." The 91-minute movie, which examines media bias during the trial, stars Fred Thompson as William Jennings Bryan, Colm Meaney as H.L. Mencken, Brian Dennehy as Clarence Darrow, and Nathan West as fictitious reporter Charles Anderson, who is torn between ethically reporting the trial and distorting the facts to impress Mencken, his mentor. Mencken is portrayed as an unethical, sexist reporter, while Darrow, Bryan, Dayton, and fundamentalists are treated sympathetically.

Alleged, which shows Scopes coaching, links evolution with eugenics through George Hunter's *A Civic Biology* (the textbook used by Scopes) and Abigale, a "half breed" housed at the "Women's Colony for Epileptics and the Feebleminded" who is targeted for sterilization after scoring poorly on an IQ test. Darrow, who rejected eugenics, helps stop Abigale's sterilization. The movie ends with a quote from John Scopes: "There is more intolerance in higher education than in all the mountains of Tennessee" (Scopes and Presley, 1967).

Source

Scopes, John T., and James Presley. 1967. *Center of the Storm: Memoirs of John T. Scopes*. New York: Holt, Rinehart and Winston, 276.

American Civil Liberties Union (ACLU)
Est. 1920

The American Civil Liberties Union (ACLU) is a nonprofit organization founded on January 19, 1920, by a group of people that included Helen Keller, Roger Nash Baldwin, and Arthur Garfield Hays. The organization pledged "to defend and preserve the individual rights and liberties guaranteed to every person in this country by the Constitution and laws of the United States." Most of its earliest cases focused on free-speech rights of artists, unionists, Communists, and other unpopular targets. Prior to the Scopes Trial, the ACLU had amassed few victories and was largely unknown outside of New York.

In March 1925, ACLU secretary Lucile Milner noticed a newspaper article describing Tennessee's passage of the Butler Act, which banned the teaching of human evolution in the state's public schools. Milner showed the article to ACLU executive director Roger Baldwin, who decided to challenge the law. Because the ACLU had no offices in Tennessee, the ACLU's publicity director, poet and novelist Joseph Freeman, suggested that a story be sent to several Tennessee newspapers to find "a Tennessee teacher who is willing to accept our services in testing [the Butler Act] in courts. Our lawyers think a friendly test-case can be arranged without costing a teacher his or her job. Distinguished counsel have volunteered their services. All we need now is a willing client." When businessmen in Dayton convinced Scopes to be the "willing client" and test the law, trial instigator George Rappleyea telegraphed the ACLU, "Prof. J.T. Scopes, teacher of science Rhea County [sic, Central] High School, Dayton, Tenn., will be arrested [and] charged with teaching evolution. Consent of superintendent of education for test case to be defended by you. Wire me collect if you wish to cooperate and arrest will follow."

At subsequent meetings in New York after Scopes' arrest, ACLU officials made it clear that they did not want Clarence Darrow involved in the case, fearing that his "headline chasing" would detract from the trial's constitutional issues. Instead, the ACLU wanted to be represented by John Davis (the Democratic Party's nominee who had been defeated by Calvin Coolidge in 1924) and Charles Hughes (a former governor of New York and justice of the U.S. Supreme Court). Scopes, however, insisted that Darrow defend him ("I want Darrow"), and Darrow remained on the defense team throughout the case. After Darrow and Malone joined Scopes' defense, the ACLU never regained control of the case.

Some were critical of the ACLU's support of Scopes; for example, *New Republic* told its readers that the ACLU had "as much reason to affirm the liberty of the people of Tennessee to control the curriculum of their public schools as it had to affirm the liberty of Mr. Scopes to teach the doctrine of evolution in those schools.... The right of Tennessee to pass a foolish law is quite as much in need of defense as the right of a soap-box orator to utter a foolish or a dangerous speech" (The conduct of the Scopes Trial, 1925). The Scopes Trial cost the ACLU $8,993.01, most of which was spent on expert witnesses. The ACLU did not pay off its Scopes-related debt until 1926, when the American Association for the Advancement of Science (AAAS) asked its members to give money for Scopes' defense. The Scopes Trial, a public-relations success for the ACLU, led to the ACLU's first endorsement by a major U.S. newspaper, the *St. Louis Post-Dispatch* (Walker, 1990).

Today, the ACLU has more than 1,200,000 members and an operating budget exceeding $300 million. It continues to fight for civil liberties, including issues such as same-sex marriage, abortion rights, prisoners' rights, and the separation of church and state.

Sources

The conduct of the Scopes Trial. 1925. *New Republic*, 43, 331–333.
Walker, Samuel. 1990. *In Defense of America Liberties: A History of the ACLU.* Oxford: Oxford University Press, 75.

Anti-Evolution Organizations

Soon after the Scopes Trial, New York preacher John Roach Straton—claiming that God had chosen him to assume William Jennings Bryan's role as leader of the fundamentalists' antievolution crusade—announced, "Some organization" was "needed to combat the Godless theory of evolution" (Straton defends Supreme Kingdom, 1927). New organizations appeared throughout the country. Some of these organizations, such as the Research Science Bureau, were "on-paper" creations that used mailing lists and pamphlets to raise money for preachers (e.g., Harry Rimmer), while other organizations boasted tens of thousands—even millions—of members and became powerful social forces. The most influential of these organizations in the 1920s was William Bell Riley's World's Christian Fundamentals Association (WCFA), which recruited Bryan to represent its interests at the Scopes Trial. Most anti-evolution organizations founded around the time of the Scopes Trial were short-lived.

American Anti-Evolution Association was formed in 1927 by Arkansas preacher Ben Bogard to help gather signatures to force a referendum on banning the teaching of human evolution in Arkansas's public schools. Late the next year, the referendum passed by popular vote (108,991 to 63,405), making Arkansas the third (with Tennessee and Mississippi), and last, state to criminalize the teaching of evolution. This law, which would outlive Tennessee's Butler Act, was the only antievolution law passed by a popular vote. After the vote, however, the organization had little impact and soon disappeared.

Anti-Evolution League of America was founded by William Bell Riley in 1924, one year after he founded the Anti-Evolution League of Minnesota. William Jennings Bryan, a friend of Riley's, supported the League's work. The league's first president was Kentucky preacher J.W. Porter, who helped present an anti-evolution bill to the Kentucky legislature. The League's field-secretary, T.T. Martin, edited the organization's newspaper (*The Conflict*) and preached tirelessly that "evolution sweeps away not only God's word, but also takes away the Savior as our Redeemer," while condemning scientists who had "captured our tax-supported school system and are trampling on our religious liberties" (Evangelist flays evolution, 1924). In 1926, Martin—who was prominent in Dayton throughout the Scopes Trial—was instrumental in convincing the Mississippi legislature to criminalize the teaching of evolution. The League organized a national "Bible-and-Christ-and-Constitution Campaign against Evolution in Tax-Supported Schools" and offered its presidency to Bryan's son, William Jennings Bryan, Jr. (Anti-evolutionists ask Bryan, Jr., to become president, 1925). After achieving a few minor victories, the League—lacking dynamic leadership—disappeared as the public lost interest in the evolution controversy.

Bible Crusaders of America was founded in November 1925 in Clearwater, Florida, to "combat Modernism, Evolution, Agnosticism, and Atheism at every point," remove evolution from public schools, and prevent "our becoming a pagan nation." The Crusaders were founded and financed by real estate tycoon George Washburn, a friend and business partner of Bryan who "dedicated his life and his fortune to war on the teaching of evolution" and saving "the orthodox Bible from the onslaught of modernism" so that "we will not have a church without Christ, a pagan country without a Bible, and a humanity without a God" (His wealth to carry on, 1926; Bible crusader, 1926). (Washburn also pledged $25,000 to create William Jennings Bryan University, which is now Bryan College.) Washburn, who operated the Crusaders' radio station and magazine (*The Crusaders' Champion*), recruited Straton, Riley, Martin, and other fundamentalist leaders to promote the Crusaders' agenda. At its first conference, which was held in Atlanta, the Crusaders pledged to make Christianity the official religion of the United States. Attendees heard T.T. Martin declare war on evolution because aside from the fall of Adam and Eve, the teaching of evolution in tax-supported schools was the greatest curse ever witnessed on Earth. The Crusaders were important in the passage of Mississippi's ban on the teaching of evolution in 1926; their slogan became "Back to Christ, to the Bible, and to the Constitution" as they promised that their "battle will never end until every evolutionist is driven from the tax-supported schools of America" (Allen, 1926). However, the Crusaders disappeared in the late 1920s when Florida's real-estate boom crashed and took most of Washburn's wealth. The October 1926 issue of *The Crusaders' Champion* was the organization's last.

Bryan Bible League was founded in California in August 1925 by evangelist Paul Rood, who claimed that God had told him to form the group to continue Bryan's fight against the teaching of evolution (Anti-evolution league formed at big meeting, 1925). The League, which eulogized Bryan as "The Modern Elijah," was endorsed by Bryan's widow, Mary Baird Bryan. More than 1,000 people joined at the group's first meeting, at which Rood promised "we are after a million members." Rood later announced that "the moral and religious effects of evolution are too disastrous for us to be silent.... Patriotism demands it. The fight is on!" (Converting California, 1926). When the League—one of the few anti-evolution organizations on the West Coast—pledged to introduce an antievolution law in 1927, newspapers reported the story with the headline "Fat Chance" (Fat chance, 1925).

Defenders of the Christian Faith was formed in November 1926 by 26-year-old Gerald Burton Winrod (1900–1957).[1] Initially, the Defenders focused on banning the teaching of evolution, but they soon also opposed homosexuality and supported Prohibition and racial segregation, all in hope of saving as many as possible "from Satan's grip." Within a few years, Winrod's organization, which was headquartered in Wichita, Kansas, touted 3,000 members and more than 10,000 subscribers to its magazine, *The Defender*. Winrod, with the help of other fundamentalists such as William Bell Riley, organized the "Flying Fundamentalists," a group that—following the model used by similarly named Prohibition activists—crisscrossed the country promoting their agenda. Winrod claimed, "I would rather my babies' eyes be gouged out at this minute than to have them taught this blatant atheism" of evolution.

By 1929, the Defenders were promising attendees at their convention that "unpleasant surprises are in store for those who believe in the affinity of Man and monkey" (Faith defenders of U.S. to meet, 1929) as they condemned Communism, socialism, evolution, and "godless humanism." Winrod, an anti-Semite who believed that the United States was the chosen land of God, later proclaimed that Hitler would save Europe from Communism, ran unsuccessfully for the U.S. Senate in 1938, and was charged (but never convicted) of sedition in 1942. As was true for the other Scopes-inspired antievolution groups, the Defenders' influence soon began to fade. Remnants remained in Kansas until at least the late 1970s (some members also touted their membership in the Christian Identity Movement), but they registered no significant accomplishments.

The Supreme Kingdom was organized in Atlanta in January 1926 by Edward Young Clarke, a political organizer turned marketing entrepreneur who had recently been kicked out of the Ku Klux Klan for violating the Mann Act. Clarke's organization was dedicated to ridding public schools of the teaching of evolution, which Clarke believed had caused a "crime wave" and "the very foundations of liberty, morals and Christianity to totter." In place of evolution, Clarke promised to "rebuild in the minds of our children the religion of our fathers" by opening offices throughout the country that would examine teachers, reject "tainted" books, and control the content of newspapers and movies (Allen, 1926). Clarke even promised to create a home in Florida "for those who grow old in the war against evolution" (Shipley, 1927). Evolutionist Maynard Shipley, who assisted Scopes' defenders in Dayton, claimed that The Supreme Kingdom threatened to use "damnable and detestable books on evolution" to fuel "bonfires" across the nation, and that the group heralded "the overthrow of our present form of government" (Shipley, 1927). Clarke hired New York fundamentalist John Roach Straton to give antievolution speeches for the Kingdom's "aggressive warfare" against evolution. (Straton's pay for 60 speeches was to be $30,000, which dwarfed his $12,000 annual salary at Calvary Baptist Church.) Clarke promised that by 1928, "from Maine to California, and from the Great Lakes to the Gulf, there will be lighted in this country countless bonfires, devouring those damnable and detestable books on evolution" (Shipley, 1927). However, in early 1927—soon after the Kingdom announced that it had chapters in 16 states—the *Macon Telegraph* reported that Clarke was keeping 50 percent to 75 percent of the membership fees for himself (in addition to his $25,000 annual salary). Days later, after denouncing the story as "unfounded," Straton resigned from the Kingdom. Early the following year, Clarke tried to recoup his losses by welcoming Catholics (who were excluded by the KKK) into the Kingdom—provided they were White, Christian, and had $10—but the Kingdom soon faded away (Macon newspaper "calls" Straton, 1927).

To combat these anti-evolution organizations, the Science League of America was "forced into existence" by Kirtley Mather. The League's members included several prominent scientists (e.g., Luther Burbank, David Starr Jordan), who believed that anti-evolution laws "would go far toward plunging us backward into the abyss of the Dark Ages" (Cardiff, 1924). In 1927, recognizing that fundamentalists could not be laughed out of existence, Mather documented the Science League's concerns in *The War on Modern Science*.

Today, anti-evolution behemoths such as the Institute for Creation Research (established 1970) and Answers in Genesis (established 1994) continue to condemn the teaching of evolution. All of these organizations cite the Scopes Trial as a critical event in the history of the evolution controversy.

Sources

Allen, Harbor. 1926. The anti-evolution campaign in America. *Current History*, 24 (6), 893–897.
Anti-evolution league formed at big meeting. 1925. *Modesto (CA) Bee*, August 14, 3.
Anti-evolutionists ask Bryan, Jr., to become president. 1925. *Arizona Republican*, September 6, 6.
Bible crusader. 1926. *Word and Way* (Kansas City, MO), January 21, 2.
Bryan's benediction. 1925. *The King's Business*, 16 (2), 535.
Cardiff, I.D. 1924. The Science League of America. *Science*, 62, 111.
Converting California. 1926. *Baltimore Evening Sun*, July 9, 21.
Evangelist flays evolutionary theory. 1924. *Shreveport Journal*, December 1, 13.
Faith defenders of U.S. to meet. 1929. *Pittsburgh Press*, March 30, 5.
Fat chance. 1925. *Arroyo Grande Valley (CA) Herald Recorder*, October 22, 2.
His wealth to carry on for W.J. Bryan. 1926. *Healdsburg (CA) Tribune*, April 17, 6.
Macon newspaper "calls" Straton. 1927. *The Sunday Record* (Columbia, SC), January 9, 1, 10.
Shipley, Maynard. 1927. *The War on Modern Science: A Short History of the Fundamentalists' Attacks on Evolution and Modernism*. New York: Alfred A. Knopf, 47, 48, 485.
Straton defends Supreme Kingdom. 1927. *Atlanta Constitution*, January 8, 18.

Brewer, Thomas Jefferson
1920–2003

Thomas Jefferson "Tommy" Brewer was born in Dayton, Tennessee, in 1920, and lived across the street from the Rhea County Courthouse on Third Avenue. On the first day of the trial, Tommy's parents took him to the courtroom, where he sat on Judge Raulston's desk and drew the names of 20 potential jurors from a hat. For his work, Tommy was paid a silver dollar.

When trial instigator George Rappleyea left Dayton after the Scopes Trial, Tommy's father, Walter E. Brewer, became manager of the Cumberland Coal and Iron Company. In 1927, Tommy's family moved into the Mansion, which had housed the previous managers of the Dayton Coal and Iron Company (DCIC). After graduating from Rhea Central High School, Tommy served aboard USS *Butte* during World War II. After the war he worked as a mechanic in Chattanooga, and he retired from American Manufacturing Company in 1985 (Thomas Jefferson Brewer, 2003). Late in his life, Tommy admitted that he remembered little about the trial except that the crowds "made it difficult to get to the popcorn stand."

Brewer, the last living participant of the Scopes Trial, died on August 16, 2003, in Red Bank, Tennessee. He was buried in Hamilton Memorial Gardens in Hixon, Tennessee.

On the first day of the trial (July 10, 1925), four-year-old Thomas Jefferson "Tommy" Brewer sat on Judge Raulston's desk and drew the names of jurors out of a hat. Brewer, who later recalled that the trial's crowds made it hard for him to get to the popcorn stands was paid a silver dollar for his work (RCHGS).

Source

Thomas Jefferson Brewer, 82, of Red Bank, died Saturday. 2003. *The Chattanoogan*, August 16, 9.

Bryan College Est. 1930

In 1905, William Jennings Bryan had lamented to his friend Evelyn McClusky, "People are not paying attention to the Bible. I wish for a college that would present Christ first" (Olson, 1995). To help make that happen, and on the day before he left Miami for Dayton, Bryan signed his will, whose eleventh bequest set aside $50,000 "for Christian education."

During the first weekend of the Scopes Trial, Henry Rusby—dean of the pharmacy college at Columbia University—denounced the Butler Act as "a reversion to slavery," "an evil policy," and "a gross offense against civilization," and recommended that his and other schools reject graduates of Tennessee's public schools (Advise schools, 1925).[2] When he heard Rusby's widely publicized claims, Rhea County School Superintendent Walter White announced that Dayton would build its own fundamentalist college, name it after Bryan, and invite Bryan to be its president (Giddens, 2007; Olson, 1995). That university is now Bryan College.

On the Thursday after the trial (i.e., July 23), Bryan toured Dayton with some friends. When he climbed nearby Matthews Hill to overlook town, he commented, "What a wonderful spot for a Christian school." Bryan envisioned a school for boys that spanned from their junior year of high school through the sophomore year of college, which he believed was the most important time for young Christians' spiritual development. Bryan planned to meet with the group to discuss the school on the following Monday, but he died the day before, Sunday, July 26, in Dayton.

On August 12, at a meeting of the Dayton Progressive Club, one of the Scopes Trial instigators, F.E. Robinson, organized the Bryan Memorial University Association (BMUA) to begin planning a school to honor Bryan.[3] Much of the initial meeting was devoted to condemning Darrow. There was also an immediate conflict, for Tom

Stewart, John Raulston, and others from the Civitan Club in Winchester, Tennessee, who claimed that Bryan had told them that he wanted a boys' school in Dayton, and a girls' school in Winchester. After the meeting, the Winchester group bowed out, but the question lingered: Where should the university be?

Backers such as William Bell Riley and the World's Christian Fundamentals Association (WCFA) wanted the university to be in Chicago and associated with Wheaton College; others wanted it in Dallas or Miami, and Dayton's leaders wanted it in Dayton. To resolve the issue, Robinson went to Miami on August 21, 1925, to meet with Bryan's widow, Mary, and their son, William Jennings Bryan, Jr., who told Robinson, "There could be no more beautiful or more fitting place for [William Jennings Bryan University] than in Dayton, where his last days were spent" (Bryan family approves, 1925).

On October 10, 1925, the Tennessee Secretary of State chartered the BMUA, whose members included prominent politicians (e.g., Georgia Congressman William D. Upshaw) and fundamentalists (e.g., New York preacher John Roach Straton). The group's 52-member National Campaign Committee, which included Tennessee governor Austin Peay, hoped to raise $5 million from 50,000 people each pledging $100, as described by the campaign's alliterative slogan, "Fifty Thousand Fundamentalists for the Faith of Our Fathers" (Bryan College, 2020). Half of the money was to create an endowment, and half was for equipment and construction of the school's 14 buildings, which included two science buildings. In Dayton, fundraising was headed by W.C. Bailey, owner of Bailey Hardware; mayor A.P. Haggard and his son Wallace, who owned and operated American National Bank (Wallace also was among John Scopes' prosecutors); and Fred Knight, the new principal at Rhea Central High School. The committee also included Sue K. Hicks, Ben McKenzie, E.B. Ewing, and John L. Godsey. The first cash gift (of $1,000) came from C.A. Dagley, a Florida realtor and former resident of Rhea County. Other donations included $1,000 from F.E. Robinson, 100 books from the Rev. Charles Potter, a set of Darwin's books from Dudley Field Malone, and a ton of coal from George Rappleyea, who was described as "a rabid evolutionist and modernist" who "started this test case to make a monkey out of the legislature" (Bryan College fund growing, 1925).

The association's 12-page prospectus described William Jennings Bryan University (WJBU) as "An Institution Founded Upon the Conviction That the Bible Is the Word of God," and university officials promised that no money would be accepted from evolutionists or from anyone not endorsing the infallibility of the Bible. However, Robinson also promised that the university would "teach science, and why not? We are going to teach the theory of evolution, too—but, mind you, I say the theory. We put that question to Mrs. Bryan.... She said, 'Surely, teach science, but teach fact as fact, and theory as theory.' We haven't any quarrel with the evolutionists. If they want to believe they descended from monkeys, we don't care. But we do object to scientists presenting mere theories to our children as cold facts" (Bryan College, 2020). The proposed school was to be nondenominational, largely because its founders came from at least eight different denominations.

The BMUA campaign was launched on Friday, October 30, 1925, with a banquet in Dayton at which Straton was the featured speaker. Days later at their "Victory Dinner" at Hotel Aqua, Robinson and Bailey announced that Rhea County had pledged $202,500 for the university, which was more than double its goal. The largest contribution from Rhea County came from mayor A.P. Haggard, whose $40,000 was earmarked for the school's Rena Clark Haggard Memorial Hospital (honoring Haggard's late wife), for which Dayton agreed to provide another $30,000. When five farmers pledged land for the proposed school, a committee examined the sites and decided to locate the school on 81 acres atop Matthews Hill. That land's owner, J.T. Matthews, sold the land to BMUA for $6,654.33, which was half of its appraised value. The first building set for construction was the 70,000 ft², three-story Administration Building, whose 440' × 54' dimensions were modeled on the biblical dimensions of Noah's Ark (Giddens, 2007). By June 14, 1925, pledges exceeded $425,000, and by July 5, 1928, totaled more than $750,000. These pledges came from 29 states, despite there being campaigns in only five.

On November 5, 1926, before a crowd of more than 1,000 onlookers, Tennessee Governor Austin Peay turned over the first shovel of dirt at the future site of WJBU as he repeated his plea for more Christian teachers so "There would be no conflict between science and religion in education." According to Peay, the university would "bear a magic name—a name that for more than a quarter of a century was a household word with the world.... Fortunate will be your university to bear the name of Bryan" (Livesay, 2020). In the crowd listening to Peay were Herbert Hicks, John Neal, Gordon McKenzie, and Ben McKenzie. Walter White, who had lost the race for governor to Peay three days earlier, did not attend.[4] Matthews Hill was soon renamed "Bryan Hill."

The financial pledges for Bryan's school included $100,000 from Edward Young Clarke, a former leader of the Ku Klux Klan and the founder of the Supreme Kingdom, an organization dedicated to eliminating the teaching of evolution in schools. Other notable donors included Franklin Delano Roosevelt, Mrs. Thomas A. Edison, and Bryan's widow, Mary Baird Bryan, who pledged money, artifacts, and more than 300 of Bryan's annotated books, including Charles Darwin's *The Descent of Man* and *On the Origin of Species*. When Congressman William Upshaw (D-GA) came to Dayton to officiate the pouring of the Administration Building's foundation, prospects for the university were bright.

Then came disaster. In October 1929, the financial excesses of the Roaring Twenties crashed the stock market; more than 10,000 banks failed as the Great Depression crippled the United States. The many unfulfilled financial pledges for Bryan's school became good wishes, fundraising ended, and construction stopped. The Administration Building would not be completely enclosed until 1952 (Giddens, 2007).

Despite these challenges, Robinson pushed ahead, and on July 9, 1930, he announced that the university's first president would be George L. Guille, a former Presbyterian preacher and a noted Bible authority from Athens, Tennessee. Robinson promised "to make this school one of the highest type in all lines of university education" (Bryan Memorial University opens, 1930).

On September 18, 1930, opening ceremonies for WJBU were held in the Rhea County Courthouse in the same courtroom in which John Scopes had been convicted five years earlier. Because there were no completed buildings atop Bryan Hill, the university leased the empty Rhea Central High School building (a new building for the school had been built nearby)—in which Scopes had allegedly taught evolution—for its classes. Classes began on September 30, 1930, with Amy Francis Louis Cartright, who graduated from Rhea Central High School just weeks earlier, being the first student to enroll. Amy was also the granddaughter of the Rev. Lemuel Cartright, who had given the Scopes Trial's opening prayer. In its first quarter, the university enrolled 27 students (and 74 in its second). WJBU's 1931–1932 catalog described the school as "an Institution Which Recognizes Revelation and Accepts the Supernatural" (Livesay, 2020).

In the spring of 1934, the first graduating class of WJBU was addressed by Bryan's friend, the Rev. Timothy Callahan, who claimed that Scopes and Rappleyea were part of a Communist conspiracy. Months later, frustrated students begged acting-president Judson Rudd to begin offering classes in the partially finished Administration Building. Rudd agreed, and students and faculty began converting seven of the building's unfinished, first-floor rooms into classrooms. In fall 1935, classes began on campus, where they have met ever since.

WJBU grew slowly, but steadily. In 1958, the school's name was changed to William Jennings Bryan College, and in 1993 it was shortened to Bryan College. In February 2000, the Administration Building—now called Mercer Hall—burned, but was subsequently restored. In 1989, Bryan College opened its Center for Origins Research (CORE) to develop and teach a creationist model for biological origins. However, in 2014, President Stephen Livesay—citing budget shortfalls—closed the center.

In February 2014, Livesay added to Bryan College's 80-year-old "Statement of Belief" that Adam and Eve "are historical persons created by God in a special formative act, and not from previously existing life-forms." Hundreds of students protested, several professors resigned, and two tenured professors were fired when they would not endorse the new statement. The faculty then voted 30 to 2 "no confidence" in Livesay.

Today, Bryan College is a private, fully accredited, liberal arts college covering 128 acres atop Bryan Hill that enrolls more than 1,400 students in more than 30 accredited majors.

Bryan College, established in 1930 in the wake of the Scopes Trial, is the most significant product of the Scopes Trial and the most enduring memorial to William Jennings Bryan and his ideals (Randy Moore).

Sources

Advise schools refuses to take Tennessee credits. 1925. *Herald and Review* (Decatur, IL), July 12, 1.
Bryan College. 2020. Establishing Bryan College. Accessed October 1, 2020, at https://archive.org/stream/EstablishingBryanCollege/Establishing%20Bryan%20College_djvu.tx
______. 2021. *Catalog 2020–2021.* Dayton, TN: Bryan College, 8.
Bryan College fund growing. 1925. *Knoxville News-Sentinel*, July 17, 1.
Bryan family approves Dayton as memorial site. 1925. *Dayton Herald*, August 27, 1.
Bryan Memorial University opens first term Thursday. *Chattanooga Times*, September 14, 1.
Giddens, Stephen D. (Editor). 2007. *Lions' Pride.* Nashville: The Booksmith Group.
Livesay, Corinne. 2020. *William Jennings Bryan: A Difference Maker.* Dayton, TN: Bryan College.
Olson, LaDonna Robinson. 1995. *Legacy of Faith: The Story of Bryan College.* Dayton, TN: Bryan College, 3.

Bryan, Mary Baird
1861–1930

Mary Elizabeth Baird was born on June 17, 1861, in Perry, Illinois, the only child of merchant John Baird (1823–1905) and his wife Lovina (1828–1896) Baird. At age 18, Mary attended Jacksonville Female Academy (which local men dubbed "Jail of the Angels"), from which she graduated as valedictorian. While there, she met and began dating William Jennings Bryan. The couple married on October 1, 1884, after which they moved to Jacksonville, Illinois, where William began practicing law. The couple had three children: Ruth Baird Bryan (1885–1954), William Jennings Bryan, Jr. (1889–1978), and Grace Dexter Bryan (1891–1945). Mary and William also took in and paid for the education of Yaschichira Bryan, a 19-year-old Japanese student who returned to Japan in 1903.

After college, Mary devoted her life to her husband and various political causes. For example, in one of William's campaigns for president, she organized the first voter-database and took classes in German so she could read articles from Europe about her husband and the economy. Mary also studied law and—after moving to Nebraska in 1888—was admitted to the Nebraska bar so she could help her husband. Mary often accompanied her husband while he campaigned for president, and later wrote the best-selling *The First Battle: A Story of the Campaign of 1896*, which included Bryan's speeches and other materials from the 1896 presidential election. For many years, Mary and William Bryan shared a large double desk so that they could work across from each other.

Throughout most of their marriage, Mary handled her husband's correspondence, negotiated with his friends and colleagues, and edited and compiled research for his speeches, articles, and books. When William began publishing his popular *The Commoner* in January 1901, Mary handled editorial duties when her husband was away. She often traveled and attended political events with her husband. After a world tour with William in 1905, one newspaper noted that Mary "has probably traveled over more countries and met more kinds of people than any other woman in the United States" (Cole, 2021).

Mary was also a leader in the Woman's Christian Temperance Union and the National American Woman Suffrage Association. Mary, who was described as "a thoroughly feminine woman with the mind of a thoroughly masculine man" (Livesay, 2020), promoted woman suffrage throughout Florida, including in an address to the Florida legislature. In her speeches, Mary promised that giving a woman the right to vote would not "unsex her [or] make her coarse or vulgar" (Mrs. William Jennings Bryan, 1917).

Throughout his illustrious career, William Jennings Bryan attributed his many successes to his wife, Mary. This photo shows the Bryans in Miami, Florida, in 1925, not long before they left for John Scopes' trial in Dayton (*RCHGS*).

Mary opposed her husband's involvement in the Scopes Trial; as her daughter Grace later told Scopes' prosecutor Sue Hicks, Mary "did all she could to prevent father from taking part in the Scopes Trial." However, once William agreed to get involved, Grace admitted, "It was, however, [Mary's] greatest comfort that her wifely devotion finally had triumphed over her opinions" (Bryan to Hicks, 1940). Mary was concerned about her husband, and at one point during the trial noted that he looked like "an old war-horse," adding, "I never saw him quite so agitated. He trembled when he stood up" (Farrell, 2011). Through it all, Mary's husband understood and appreciated the many ways that Mary helped him, often telling people, "Whatever success I have attained I owe to my wife" (Mrs. Mary Baird Bryan, 1930).

Despite being confined to a wheelchair, Mary attended most sessions of the Scopes Trial. Mary believed, "Scopes violated the law, and the efforts of the opposition to make the case hinge on the truth or lack of truth in the theory of evolution were out of place.... It was irrelevant to the subject when men far from unblemished reputation exclaimed, 'I am a Christian; just as good a Christian as Mr. Bryan'" (Bryan and Bryan, 1925). Mary said little to the press during the trial, but "bulletins" to her absent daughters described the trial's first week as "a great mixture of waving fans, perspiring men, and angry lawyers." Darrow had "a weary, hopeless expression ... [and] is full of sarcasm and bitterness," and Hays was "as self-asserting as the New York Jews can be ... his eyes [are] filled with shrewdness." Judge Raulston was "a pleasant-faced man ... who tries to make everybody comfortable and happy and wants to do the fair thing," and Neal "has the appearance of being suddenly frightened up out of the grass like a startled rabbit." Scopes, "'a long-jawed mountain product,' carries himself wretchedly ... his whole appearance is simpering, weak, and gawky" (Cherney, 1985; Farrell, 2011). After the trial, Mary claimed, "Mr. Scopes's defense was defeated at every point and the decision of the court [was] a triumph for the Tennessee statute" (Bryan and Bryan, 1925).

When her husband died in Dayton five days after the trial, Mary insisted on a simple ceremony, telling reporters, "We are simple people and we want all arrangements simply made." Mary met her family in Washington, D.C., for the funeral and burial at Arlington National Cemetery, instructing her children to "bring your white clothes; don't bring any black ones." She also lamented, "Oh, I so wanted him to be left to me for a few years. There are so many things that he was needed for" (Acclaim from millions pours in, 1925).

After the funeral, Mary returned to Miami and—telling people that "the world needs his memoirs"—completed her husband's memoirs. In that book, Mary recognized "just why the interest [in the evolution controversy] grew, just how he was able to put fresh interest into a question that was popular twenty-five years ago, I do not know ... [but] Whenever Mr. Bryan took a stand upon any subject, the matter at once became an issue. People began to fall in line. Sides grew distinct. The public divided and stood ready to do battle.... The vigor and force of the man seemed to compel attention" (Bryan and Bryan, 1925; Widow desires simple funeral, 1925).

Mary, who was often referred to as the "Commoner's Inspiration" who "helped to bring to fullest flower his gift of eloquence," died on January 21, 1930, at the home of her daughter, Grace Hargreaves, in Hollywood, California. Her ashes were taken to Virginia by her son, William Jennings Bryan, Jr., where they were interred beside her husband in a private ceremony at Arlington National Cemetery as her husband's "wife and helpmate" (Livesay, 2020; Death takes Bryan's widow, 1930). In its obituary for Mary, the *Miami Daily News* told its readers, "The world is better for Mrs. Mary Baird Bryan having lived in it" (Mrs. Mary Baird Bryan, 1930).

Sources

Acclaim from millions pours in on Dayton for noted dead. *Selma (AL) Times*, July 28, 1–7.
Bryan, Grace Dexter, to Judge Sue K. Hicks. April 12, 1940. Sue K. Hicks Papers (MS 1018), University of Tennessee, Knoxville.

Bryan, William Jennings, and Mary Baird Bryan. 1925. *The Memoirs of William Jennings Bryan*. Philadelphia: John C. Winston, 479–484.
Cherney, Robert W. 1985. *A Righteous Cause: The Life of William Jennings Bryan*. Boston: Little, Brown, 176.
Cole, Nancy. 2021. Biographical sketch of Mary Baird Bryan. In Part III: Mainstream Suffragists—National American Woman Suffrage Association. Accessed September 2, 2021, at https://documents.alexanderstreet.com/d/1009859961.
Death takes Bryan's widow. 1930. *Palm Beach (FL) Post*, January 22, 1.
Farrell, John A. 2011. *Clarence Darrow: Attorney for the Damned*. New York: Doubleday, 374–382.
Livesay, Corinne. 2020. *William Jennings Bryan: A Difference Maker*. Dayton, TN: Bryan College, 19, 31, 84
Mrs. Mary Baird Bryan. 1930. *Miami Daily News*, January 22, 6.
Mrs. William J. Bryan Gives Suffrage Talk. 1917. *Daytona (FL) Daily News*, January 26, 5.
Widow desires simple funeral. 1925. *Nashville Banner*, July 27, 1.

Bryan, William Jennings
1860–1925

William Jennings Bryan was born into a rich, religious family on March 19, 1860, in Salem, Illinois. After becoming a Christian at age 14 and graduating atop his class from Illinois College in 1881, he earned a LL.B. from Chicago's Union Law College (now Northwestern University School of Law) two years later. In 1884, Bryan married Mary Elizabeth Baird, who he had met five years earlier in college. Throughout their marriage, Mary helped her husband in countless ways, including managing his correspondence, editing his publications, and fine-tuning his speeches.

Throughout the trial, Scopes' prosecutor William Jennings Bryan often cooled himself with a fan that advertised Robinson's Drug Store, where the trial originated (RCHGS).

In 1887, the Bryans moved to Lincoln, Nebraska, from which he served two terms as a Democrat in the U.S. House of Representatives beginning in 1891.[5] After a failed Senate bid in 1894, Bryan edited the *Omaha World-Herald* while remaining active in Democrat politics, and in July 1896 attended the Democratic Party convention in Chicago. There, the 36-year-old Bryan—little-known outside Nebraska—delivered his electrifying "Cross of Gold" speech attacking the gold standard and promoting the increased coinage of silver. His speech, which recalled the crucifixion of Jesus Christ, concluded with "We will answer demands for a gold standard by saying to them: 'You should not press down upon the brow of labor this crown of thorns, you shall not crucify mankind on a cross of gold.'" The speech catapulted Bryan "into the headlines of history" (Kazin, 2006) and secured for Bryan the Party's nomination for president, after which he travelled extensively while inventing the "stump tour" that is common today. "Boy Bryan" remains the youngest nominee of a major political party—and the youngest to receive an electoral vote—in U.S. history.

In the 1896 presidential election, Bryan pioneered campaigning by car and train, traveling more than 18,000 miles across 26 states to deliver 600 speeches promoting the free coinage of silver. Bryan's opponent, Republican William McKinley, promoted the gold standard, which Bryan denounced as anti-American (Kazin, 2006). Facing an almost 40-fold deficit in campaign contributions, Bryan lost the election (271 to 176 of Electoral College votes, 51 percent to 47 percent of the popular vote). After losing again to McKinley in 1900 (292 to 155 of Electoral College votes, 52 percent to 46 percent of the popular vote), Bryan organized the 2,000-member Nebraska National Guard for the Spanish-American War, but he never served in combat.

In 1901, Bryan—dubbed "The Commoner" by reporter Willis J. Abbot because of his faith in the innate goodness and wisdom of common people (Kazin, 2006)—turned his attention to writing and speaking. In January of that year, he published the first of 768 issues of *The Commoner*, a popular newspaper whose circulation eventually exceeded 250,000 subscribers. In *The Commoner*, Bryan promoted his Christian populism; he seldom separated politics and religion, often describing his political policies as "applied Christianity." In 1904, Bryan delivered his first major (and most famous) speech condemning evolution. Bryan gave several versions of that speech, titled "The Prince of Peace," dozens of times during the last two decades of his life.

In 1908, after again becoming the Democratic Party's nominee for president, Bryan used "Shall the People Rule?" as his campaign's slogan. Bryan, who opposed U.S. imperialism while promoting the federal regulation of banks and support for highways and education, was defeated handily by William Howard Taft (321 to 162 of Electoral College votes, 52 percent to 41 percent of the popular vote).[6]

Bryan, by now something of a folk-hero in rural America, then turned to speaking and writing. He wrote his own speeches and said, "I could move [the audience] as I chose. I have more than usual power as a speaker.... God grant I may use it wisely." By the time his political life ended, Bryan was the most popular speaker on the Chautauqua circuit, which provided a major source of his income; he received $250 per speech, plus half of all revenues above $500, and was "good for forty acres of packed Fords, anywhere, at any time of the day or night" (Livesay, 2020). John Scopes, who knew Bryan before either came to Dayton, knew that "I probably would not ever hear another man such as he. He was the most outstanding public speaker this country has produced. With unlimited access to television so that he could have gone into every American home in 1896.... Bryan would have been elected ... [his] passionate, convincing oratory ... would have stampeded the nation to him" (Scopes and Presley, 1967; John T. Scopes to attend Bryan dedication, 1961). Near the end of his life, including while he was in Dayton, Bryan also spent much time—and earned much money—promoting real estate in Florida (Kazin, 2006).

After winning the presidency in 1912, Woodrow Wilson named Bryan the 41st Secretary of State, at which time Bryan negotiated reconciliatory "cooling-off period" peace-treaties and promised, "There will be no war while I am Secretary of State." Bryan remained in that job until June 9, 1915, when he resigned to protest Wilson's treatment of the sinking of the British passenger liner *Lusitania* (killing 1,198 people, including 128 U.S. citizens) and America's increasing belligerence toward Germany. Out of office, Bryan continued to promote progressive causes, including an eight-hour workday, minimum wage, and women's suffrage. At the 1924 Democratic national convention, Bryan helped defeat a resolution condemning the Ku Klux Klan. By that time, Bryan was also preaching against evolution in his "Tourist Bible Class" at Miami's First Presbyterian Church, which attracted about 4,000 people every Sunday. Bryan dominated the Democratic Party for 16 years (1896–1912), despite not holding an elected office during that time.

Bryan had earlier studied evolution and decided "to have nothing to do with it." He was troubled by human evolution but dismissed the rest because "a hypothesis in regard to the rocks and plant life does not affect the philosophy upon which one's life is built." But by the early 1920s, Bryan's patience with evolution had disappeared and he now considered it to be "the most paralyzing influence with which civilization has had to contend." Bryan was especially concerned about theistic evolutionists—that is, Christians who accept evolution—who were "doing more harm than the atheists" because their claims were an "anesthetic that deadens the Christian's pain while his religion is being removed" (Bryan, 1924). Bryan objected to the teaching of evolution because it disputed the truth of the Bible, diverted attention from societal problems, numbed people's spiritual lives, and paralyzed hope for reform. To Bryan, "the barbarous sentiment that runs through evolution ... dwarfs the moral nature of those who become obsessed with it.... All of the sympathetic activities of civilized societies are condemned" (Livesay 2020; Bryan, 1925). Bryan's pleas found a large, receptive audience.

In early 1923, Bryan recognized that "my power in politics is not what it used to be ... [but] it has increased in religious matters" (Cherney, 1985). He soon began telling people, "There has not been a reform for 25 years that I did not support," then quickly adding, "I am now engaged in the biggest reform of my life. I am trying to save the Christian Church from those who are trying to destroy her faith" (Gould, 1996). To Bryan, Darwinism was an unsubstantiated "hypothesis" that caused people to lose touch with God. Bryan believed that German philosopher Friedrich Nietzsche's (1844–1900) critiques of religion and Christian morality were "the ripened fruit of Darwinism," adding, "A tree is known by its fruit" (Bryan, 1922).

Bryan strongly defended the right of parents to determine what schools teach and claimed that modernists were trying to invalidate "every moral standard that the Bible gives us" (Kazin, 2006). Bryan, who claimed, "Scratch a critic of the Bible and you are sure to find an evolutionist," blamed evolution for World War I as he equated Darwinism with the might-makes-right Social Darwinism advocated by imperialists, German militarists, and capitalists such as Andrew Carnegie. Bryan was also concerned that the teaching of evolution would corrupt children, and soon joined the fundamentalists' growing antievolution campaign. Earlier in his career, Bryan's most important causes were the economy, peace, prohibition, and women's suffrage. By 1923, though, he was telling reporters, "I have the most intolerant and vindictive enemies I have ever met and I have the largest majority on my side I have ever had and I am discussing the greatest issue I have ever discussed" (James, 1997).

In 1924 in Nashville, Bryan delivered his famed "Is the Bible True?" speech, which was distributed by attorney W.B. Marr to state legislators and other dignitaries. Tennessee legislator John Butler read the speech and, early the next year, introduced a bill to ban the teaching of human evolution in Tennessee's public schools. When Butler's bill passed, Bryan telegrammed Governor Austin Peay, "The Christian parents of [Tennessee] owe you a debt of gratitude for saving their children from the poisonous influence of an unproven hypothesis."

Five days after Scopes was arrested in Dayton for violating the Butler Act, fundamentalist leader William Bell Riley asked Bryan to represent the World's Christian Fundamentals Association at Scopes' trial. Bryan was not as theologically rigid as fundamentalists; for example, he did not object to teaching evolution as a theory, and he did not want violators of antievolution laws to be penalized, because "we are dealing with an educated class that is supposed to respect the law" (Bryan and Bryan, 1925). Nevertheless, he agreed to represent Riley's organization, after which Clarence Darrow—intent on exposing Bryan's "fool ideas"—volunteered to defend Scopes.

As the trial approached, Bryan proclaimed evolution "the greatest menace to civilization as well as to religion," but added that the trial's "real issue" was "the right of the people speaking through the legislature to control the schools which they create and support ... just as a painter uses the colors that his employer desires" (Cherney, 1985). Bryan demanded, "Teachers in public schools must teach what the taxpayers desire taught—the hand that writes the paycheck rules the school" (Bryan, 1924). Bryan continued to condemn evolution, adding that he had "uncovered a conspiracy against the Bible.... The evolutionists bring their doctrine before the public in a jeweled case and praise it as if it were a sacred thing.... The evolutionists have not been honest with the public.... Evolution leads logically to agnosticism. The presence of Mr. Darrow [at the trial] ... represents the most militant anti-Christian sentiment in the country" (Is conspiracy against Bible, Bryan asserts, 1925; Bryan, 1924). Rhea County School Superintendent Walter White described the upcoming trial as "the battle of the century" as Bryan claimed that it was "better to give up schools than the Bible" (Better give up schools, 1925). Like other modernists, playwright George Bernard Shaw disagreed, telling reporters, "What [Bryan] calls fundamentalism I call infantilism" (Shaw calls ideas of Bryan "infantilism," 1925).

Bryan got to Dayton on the afternoon of Tuesday, July 7, aboard the Royal Palm Express train from Miami, where he told an awaiting crowd, "This is the day I have been waiting for" (McRae, 2009). Bryan was exceedingly popular in Dayton; the applause that greeted him when he got to town convinced Darrow that Bryan was "the ruler of the Bible belt" (Darrow, 1932). When Walter White introduced Bryan as "the world's greatest man," a reporter qualified White's praise by admitting that "when the three-ring circus opens tomorrow, in the first ring will be Mr. Bryan. In the second ring will be Mr. Bryan, and Mr. Bryan will be in the third ring. Mr. Bryan occupies the center of the stage and all the corner.... Mr. Bryan may not be the greatest man in the world, but he's the greatest man in Rhea County now" (Evolution trial opens this morning, 1925). For his part, Bryan claimed that "my fight is with the so-called 'Modernists' of the Christian Church over the matter of Christian doctrine and belief, and in this battle I am not concerned with the views of the agnostics or infidels" (Bryan, 1925). Bryan attributed the worldwide interest in the trial "to the fact that this trial uncovers an attack which a generation has been made more or less secretly upon revealed religion, that is, the Christian religion" (Dayton's amazing trial, 1925). Throughout the trial, Darrow treated Bryan—and not Scopes—as the defendant.

While visiting his parents in Paducah, Kentucky, Scopes told reporters that he'd heard that Bryan would try "to make the greatest speech of his life" at the trial (Scopes back home, 1925). Bryan had asked several leaders of the fundamentalist movement—for example, John Roach Straton, Frank Norris, and Billy Sunday—to help him in Dayton, but none showed up. In Dayton, Bryan—whom Darrow claimed had "unchangeable convictions; he did not think; he knew"—was on his own (Darrow, 1932).

Within hours after arriving in Dayton, Bryan promised that the world would witness a "duel to the death." However, when he finally rose to speak in court for the first time (i.e., on the trial's fourth day, July 16), he argued against the admissibility of the defense team's expert witnesses, complained how the textbook used by Scopes grouped humans with other mammals, emphasized that parents should be able to control what is taught in public schools, and condemned evolution. Bryan's speech elicited applause but was quickly overshadowed by Dudley Field Malone's "We Are Not Afraid" response, which Darrow described as "particularly brilliant" and which even Bryan admitted was the greatest speech he had ever heard (Livesay, 2020).

The Scopes Trial's most famous event happened on the afternoon of July 20, when the defense called Bryan to the witness stand. (Some newspapers had reported that Bryan would be a witness more than a week earlier; for example, see Kinsley, 1925.) When he was called to the stand, Attorney General Tom Stewart and the other members of the prosecution objected, but Bryan insisted that he would answer Darrow's questions, provided that Bryan would then get to question Darrow, Malone, and Hays. Darrow immediately pushed Bryan to affirm a literal view of the Bible, but Bryan resisted, responding, "Some of the Bible is given illustratively," and later adding that God "may have used language that could be understood at the time."

Throughout his confrontation with Darrow, Bryan defended his faith, admitted that he didn't know or care much about how long animals had been on Earth ("I am not so much interested in that"), and acknowledged that the Earth could be far older than the 6,000 years claimed in 1650 by Irish prelate James Ussher. Virtually all leading fundamentalists of Bryan's era accepted an old Earth, and Bryan had published his ideas about Earth's age years earlier; when charged with "believing that the world was made in six days—evidently meaning days of twenty-four hours each, I have never, in writing or in my speeches, said anything to justify such a charge. According to the interpretation placed upon it by orthodox Christians, the day mentioned in the account of creation was of indefinite duration. The only persons who talk about a twenty-four-hour day in this connection do so for the purpose of objecting to it; they build up a straw man to make the attack easier, as they do when they accuse orthodox Christians of denying the roundness of the earth, and the law of gravitation" (Mr. Bryan's rejoinder, 1923). As the questioner, Darrow had the advantage of determining the topics and terms of Bryan's testimony. Bryan never got to question Darrow, Malone, and Hays in court.

The next morning, Bryan issued a statement admitting that he was sometimes "not acquainted" with the "details" to which Darrow had referred, adding that throughout his long career he had never been called "ignorant" or "an ignoramus by anyone except an evolutionist." Fundamentalists at the time were not upset by Bryan's testimony, but reporters and many others lampooned his performance while claiming that he was defeated, embarrassed, outwitted, and "intellectually butchered" by Darrow (Hyde, 1925). *The Nation* described Bryan as a "pathetically sincere and pitifully ignorant old man," and Darrow biographer Irving Stone claimed that Bryan "had taken a terrible licking.... He had been exposed as an ignoramus with a childlike mind" (Editorial, 1925; Farrell, 2011; Stone, 1943). As noted by historian Robert Linder, "the vigorous antagonism toward Bryan displayed by American newspapers … [made Bryan] the most vilified man in America, at least in most of the nation's leading newspapers," which had misquoted and ridiculed him throughout the trial (Linder, 1975). However, Harvard law professor Alan Dershowitz later concluded that Bryan was not a "simple-minded literalist" or "the know-nothing literalist of *Inherit the Wind*. For the most part, he actually seems to have gotten the better of Clarence Darrow in the argument over the Bible" (Dershowitz, 1996). Scopes, who did not want his life to be defined by his famous trial, wrote in his memoirs that "no fair man would judge Bryan's place in history by his actions at Dayton alone; he deserves better" (Scopes and Presley, 1967).

After the trial, Bryan completed editing the speech that the trial's abrupt ending prevented him from delivering in court. While driving hundreds of miles to give several speeches, Bryan had his "last serious talk" with his wife Mary, in which she warned him about his religious zeal becoming intolerance. William asked Mary, "I have not yet made that mistake, have I?" and Mary responded, "you are all right so far, but will you be able to keep this narrow path?" When William answered, "I think I can," Mary asked if he could control his followers. Her husband again responded, "I think I can," but Mary knew that he "was adding mentally 'by the help of God'" (Bryan and Bryan, 1925).

Bryan then returned to Chattanooga, where he spent Saturday night finishing the manuscript of his undelivered speech. Bryan worked there with his friend George Milton, who later admitted that Bryan was "quivering with hurt" because Darrow's questioning had "cut Bryan to the quick" (Milton, 1925). The next morning (i.e., July 26), Bryan returned to Dayton, where he attended services (without Mary) at First United Methodist Church (South). He then went to the F.R. Rogers's residence where he and his entourage were staying, ate a large lunch, and assured his wife, "I have several more years to live" (Bryan and Bryan, 1925). He then made a few phone calls, retired to a bedroom, and died during an afternoon nap. Walter Thomison, the attending physician, listed "apoplexy" (i.e., a stroke or cerebral hemorrhage) as the cause of death.

President Calvin Coolidge ordered flags to be lowered to half-mast as civic and religious leaders eulogized Bryan as "a martyr to the faith of our fathers," "the greatest moral force of his day," "the outstanding man of our movement," "the noblest hero of these times," and "God's Napoleon … one of the world's immortals … who fell with his face to the enemy." One obituary claimed that Bryan's death was "one of the greatest losses that America has ever sustained.… There is no one to take his place," and prosecutor Ben McKenzie lamented, "True Christianity has sustained an irreparable loss" (Loop, 1925; Livesay, 2020). According to his brother Charles, Bryan had "given his life in the defense of the Christian religion." In private, Darrow told friends that Bryan "died of a busted belly," but he told reporters, "I have known Mr. Bryan since 1896 and supported him twice for the presidency. He was a man of strong convictions and always espoused his cause with ability and courage. I differed with him on many questions, but always respected his sincerity and devotion." Darrow later lamented, "I am sorry that he could not have seen all this devotion that followed him to his resting place" (Darrow praises, 1925; Darrow, 1932; Farrell, 2011; Stone, 1943). Few people had been better known, and even fewer had been more often heard, than Bryan. Bryan's death, George Rappleyea admitted to the ACLU in August, "swept away any victory we might have gained."

After a memorial service at the Rogers's home, Bryan's body was taken by train to Washington, D.C. Crowds gathered at every town and crossroad, much as they had during his presidential campaign in 1896, prompting Mary to note, "They have indeed lost a friend who never broke faith with them" (Bryan and Bryan, 1925). After more than 20,000 people filed by his casket, Bryan's funeral was broadcast nationwide on radio from Washington's New York Avenue Presbyterian Church, "The Church of the Presidents" that the Bryans had attended during the Bryan's time there. The Rev. Joseph Sizoo closed Bryan's funeral with "Thank God for the memory and heritage of William Jennings Bryan," after which Bryan was buried—as per his wishes—in Arlington National Cemetery. His epitaph came from Alexander Pope's *Moral Essays*: "Statesman, yet Friend to Truth! Of Soul Sincere, in Action Faithful, and in Honor Clear." Bryan's tombstone also declares that "He Kept the Faith" and lists his wife Mary as his "wife and helpmate." Kelso Rice, the trial's bailiff who accompanied Bryan's body from Dayton to Washington, gave the flag that covered Bryan's casket to his widow Mary. In return, Mary gave Rice the silver fountain pen used by her husband until his death.

In decades since the Scopes Trial, Bryan—whose career spanned five decades, nine presidential administrations, and two wars—has often been ridiculed for his performance in Dayton, even by creationists. For example, George McCready Price lamented "poor Bryan, with his day-age theory of Genesis, [who] was so piteously overwhelmed and defeated by the battery of evolutionary scientists under the clever leadership of the atheist Clarence Darrow ... [Bryan] really didn't know a thing about the scientific aspects of the case." To Price, the Scopes Trial was a defeat for fundamentalism that had "undoubtedly [been] planned long ahead by the devil ... the general results were a triumph for the devil's cause such as the world has seldom seen before or since" (Price, 1942). The Rev. Jerry Falwell later claimed that Darrow's questioning of Bryan "proved disastrous for Bryan and Fundamentalism," and that Bryan "lost the respect of Fundamentalists when he subscribed to the idea of periods of time for creation rather than 24-hour days" (Falwell, 1981). Similarly, Henry Morris—whose *The Genesis Flood* became the founding document of modern creationism—condemned Bryan for "walking into a deadly trap" set by Darrow and the ACLU that intimidated and "catapulted" Christians "to theistic evolution." According to Morris, "the most serious mistake made by Bryan on the stand was to insist repeatedly that he had implicit confidence in the infallibility of Scripture, but then to hedge on the geological question.... [The ridicule after the Scopes Trial] proved so discouraging and intimidating to the Christian world that the whole fundamentalist movement seemed to wither away, as least as far as impact on the secular world was concerned" (Morris, 1984).

Although Bryan's populist Christianity was often criticized—for example, a popular textbook from the 1960s claimed that Bryan "had not lent the [Democrat] Party much in intellectual distinction"—Bryan supported a variety of progressive reforms, including the Sixteenth Amendment (graduated income tax), Seventeenth Amendment (direct election of U.S. Senators), Nineteenth Amendment (women's suffrage; Livesay, 2020), and the public ownership of railroads. Before the Scopes Trial began, Bryan explained that he could "be a progressive in politics and a fundamentalist in religion" because "government is manmade and therefore imperfect.... If Christ is the final word, how may anyone be progressive in religion? I am satisfied with the god we have, with the Bible, and with Christ" (Bryan, radical in politics, 1925).

Many historians consider Bryan one of the most influential progressive politicians of his era; as one concluded, "before Bryan, the Democrat Party was a conservative party of Civil War losers. After Bryan, the party was a progressive alliance of small businesses, farmers, Blacks, and blue-collar workers" (Kazin, 2006; Livesay, 2020; Douthat, 2006; Farris, 2011). Bryan viewed the Scopes Trial not as about evolution (e.g., "I am not so sure that it is involved"), but instead as an extension of the populist ideals he had championed for his entire career; just as he had condemned oppressive and exploitative financiers in 1896, so too did he attack elitist scientists who were trying to suppress the virtuous majority of people in Tennessee by forcing "irreligion"—that is, the teaching of evolution "under the guise of teaching science"—into public schools. Bryan's dedication to populism was renowned; Herbert Hoover later claimed that President Franklin D. Roosevelt's New Deal was "Bryanism under new words and methods" (Murphy, 2002), and journalist William Allen White concluded that Bryan "influenced the thinking of the American public more profoundly than any other man of his generation" (Merritt, 1925). Biologist Stephen Jay Gould acknowledged Bryan as one of American's greatest populist reformers, but added, "Fundamentalist legislation against evolution is the only truly distinctive and enduring brand that Bryan placed upon American history" (Gould, 1996). Bryan has been honored with numerous schools, museums, hospitals, libraries, parks, statutes, postage stamps, statues, and other memorials (Livesay, 2020).

On the day before he left his home in Miami for Dayton, Bryan prepared a handwritten last will and testament that pledged $50,000 "for Christian education." When the Depression wiped out much of that money, others in Dayton and elsewhere helped achieve Bryan's dream. Today, Bryan College remains the most lasting memorial to Bryan and his ideals.

Sources

"Better give up schools than Bible," Bryan says on eve of battle. 1925. *Chattanooga Daily Times*, July 10, 1.

Bryan radical in politics, standpat in religion, tells why. 1925. *Chicago Daily Tribune*, May 28, 5.

Bryan, William Jennings. 1922. *In His Image*. New York: Fleming and Revell, 124.

______. 1924. *Shall Christianity Remain Christian? Seven Questions in Dispute*. New York: Fleming H. Revell, 125, 128, 154–155.

______. 1925. *Last Message*. New York: Fleming H. Revell, 10, 57–59.

______, and Mary B. Bryan. 1925. *The Memoirs of William Jennings Bryan*. Chicago: John C. Winston, 248–249, 479–486, 490.

Cherney, Robert W. 1982. *A Righteous Cause: The Life of William Jennings Bryan*. Boston: Little, Brown, 157, 171, 175.

Darrow, Clarence. 1932. *The Story of My Life*. New York: Charles Scribner's Sons, 251, 271.

Darrow praises Bryan's courage and sincerity. 1925. *Chicago Tribune*, July 27, 1.

Dayton's amazing trial. 1925. *Literary Digest*, 86 (4), 5–7.

Dershowitz, Alan. 1996. In Darrow, Clarence. 1996 [1932]. *The Story of My Life*. New York: Da Capo Press, vii–x.

Douthat, Ross. 2006. They made America. *The Atlantic*, 60, 64.

Editorial. 1925. *The Nation*, 8, 154.

Evolution trial opens this morning; Raulston relies on divine guidance. 1925. *Chattanooga Daily Times*, July 10, 1.

Falwell, Jerry. 1981. *The Fundamentalist Phenomenon: The Resurgence of Conservative Christianity*. Garden City, NY: Doubleday, 86.

Farrell, John A. 2011. *Clarence Darrow: Attorney for the Damned*. New York: Doubleday, 397–398.

Farris, Scott. 2011. *Almost President: The Men Who Lost the Race but Changed the Nation*. Guilford, CT: Lyons Press, 75, 85.

Gould, Stephen Jay. 1996. William Jennings Bryan's last campaign. *Nebraska History* 77, 177–183. (Originally published in *Natural History*, 96 [11], November 1987.)

Hyde, Henry M. 1925. Bryan on stand, upholds Bible statements as true. *Baltimore Sun*, July 21, 1.

Is conspiracy against Bible, Bryan asserts. 1925. *Chattanooga Daily Times*, June 19, 1.

James, Gilbert. 1997. *Redeeming Culture: American Religion in an Age of Science*. Chicago: University of Chicago Press, 23.

John T. Scopes to attend Bryan dedication rites. 1961. *Shreveport Times*, August 22, 33.
Kazin, Michael. 2006. *A Godly Hero: The Life of William Jennings Bryan*. New York: Knopf, xiv, 53–58, 61, 67, 265–266, 285–288, 304.
Kinsley, Philip. 1925. Court may bar Scopes experts. *Ft. Worth Star-Telegram*, July 12, 1, 6.
Linder, Robert D. 1975. Fifty years after Scopes: Lessons to learn, a heritage to reclaim. *Christianity Today*, 19 (21), 7–10.
Livesay, Corinne. 2020. *William Jennings Bryan: A Difference Maker*. Dayton, TN: Bryan College, 24, 46, 58, 75, 88, 94, 100, 137–144, 150–156.
Loop, Sue Mills. 1925. Not a murmur passes lips of sad widow. 1925. *Chattanooga Times*, July 28, 1–2.
McRae, Donald. 2009. *The Great Trials of Clarence Darrow: The Landmark Cases of Leopold and Loeb, John T. Scopes, and Ossian Sweet*. New York: Harper Perennial, 172.
Merritt, Dixon. 1925. Bryan at sixty-five. *Outlook*, 140, 182.
Milton, George F. 1925. A Dayton postscript. *Outlook*, 140, 550–552.
Mr. Bryan's rejoinder. 1923. *The Forum*, 70 (July–December), 1852–1853.
Morris, Henry M. 1984. *A History of Modern Creationism*. San Diego: Master Books, 65–68.
Murphy, Troy A. 2002. William Jennings Bryan: Boy orator, broken man, and the "evolution" of America's public philosophy. *Great Plains Quarterly*, 22 (2), 83–98.
Price, George McCready. 1942. What Christians believe about Creation. *Bulletin of Deluge Geology*, 2, 76.
Scopes back home to see his mother, says Darrow can outwit prosecution. 1925. *Paducah Evening Sun*, July 3, 1.
Scopes, John T., and James Presley. 1967. *Center of the Storm: Memoirs of John T. Scopes*. New York: Holt, Rinehart and Winston, 209–210, 216.
Shaw calls ideas of Bryan "infantilism." 1925. *New York Times*, June 10, 1.
Stone, Irving. 1943. *Clarence Darrow for the Defense*. Garden City, NY: Doubleday, 64, 462.

Bryan, William Jennings, Jr.
1889–1978

William Jennings Bryan, Jr., was born on June 24, 1889, in Lincoln, Nebraska, and was the only son of William Jennings Bryan and Mary Baird Bryan. After graduating from the University of Arizona and Georgetown Law School, Bryan practiced law and was an assistant U.S. attorney in Arizona from 1915 to 1920. On June 17, 1925, Bryan's father urged Junior to come to Dayton for the Scopes Trial, telling him, "This trial will become one of the greatest trials in history and I want your name associated with it.... The issue will be raised in every state and I want you to be in a position to take up my work and carry it on in this matter. I can't return to the practice of law, but you can take my place in the various states.... [The trial] will give you a standing no one else can have.... Don't fail to come; it would be a very grave mistake for you to miss this opportunity" (Livesay, 2020).

At the Scopes Trial, the self-effacing Junior became friends with Scopes and fellow prosecutor Wallace Haggard, with whom he went swimming during the noon recess on the Wednesday of the trial (i.e., July 15). By the time the three returned to the courtroom, proceedings had resumed. The next day, Junior urged Judge Raulston to exclude the testimony of the expert witnesses brought to Dayton by the defense because it "is wholly immaterial, incompetent and inadmissible.... If one of the experts gives a false opinion, there is no way that you can contradict him.... [Expert testimony] is largely a field of speculation besought with pitfalls and uncertainties." Bryan added that allowing the defense team's scientists to testify would "substitute trial by experts for trial by jury" and announce Judge Raulston's "belief that this jury is too stupid to determine a simple question of fact." Throughout his argument, the soft-spoken Bryan had to be reminded to talk louder so that others could hear what he was saying. During the trial's second weekend, the younger Bryan returned to California, and therefore missed Darrow's famous interrogation of his father. Two months after his father's death in Dayton, Bryan was offered the presidency of the Anti-Evolution League of America, but the organization soon faded away (Anti-evolutionists ask Bryan Jr. to become president, 1925).

During the appeal of Scopes' conviction, Junior argued in a mailed-in statement that the Butler Act was a "deliberate, thoughtful enactment of a sovereign people, which was designed to protect their children in their own public schools in their beliefs in the divine origin of man, which in turn measures their responsibilities to God and their fellow man." Bryan, like his father, endorsed the majority's right to "protect the children of the state in the public schools in their common belief" (Supreme Court hears, 1926).

In 1938, President Franklin Roosevelt appointed Bryan the Collector of Customs for the Port of Los Angeles, where he served four terms. In 1975, Bryan moved to Santa Fe, New Mexico, to be nearer to his children. The 88-year-old Bryan died in a nursing home on May 27, 1978, in Pecos, New Mexico.

Sources

Anti-evolutionists ask Bryan, Jr., to become president. 1925. *Arizona Republican*, September 6, 6.
Livesay, Corinne. 2020. *William Jennings Bryan: A Difference Maker*. Dayton, TN: Bryan College, 114.
Supreme Court hears Scopes case. 1926. *Nashville Banner*, May 31, 1.
William Jennings Bryan, Jr., only son of the famed orator. 1978. *New York Times*, March 28, 40.

Butler, John Washington
1875–1952

John Washington Butler was born on December 17, 1875, in Macon County, Tennessee. Butler worked on his family's 120-acre corn and tobacco farm and, when he was older, taught occasionally in local schools, despite the fact that he had attended school for only 21 months. In the early 1920s, after hearing a preacher tell of a woman who rejected God after taking an evolution course at a nearby university, Butler concluded that the teaching of evolution was dangerous and could corrupt his three boys.

In 1922, Butler—a Democrat and Primitive Baptist—was elected to represent Trousdale, Macon, and Sumner counties in the Tennessee House of Representatives. Butler, who served on a committee overseeing public schools, promised voters in his rural district that he would protect children from evolution by banning its teaching in the state's public schools.

On his 49th birthday, Butler sat in his living room and drafted a bill that would make it "unlawful for any teacher" in a public school "to teach any theory that denies the story of the Divine creation of man as taught in the Bible, and to teach instead that man has descended from a lower order of animals." According to Butler, "the evolutionists against whom the law is directed deny the immortality of the soul, the virgin birth of Christ, the resurrection of the body, and that the Bible is the inspired word of God. If we are to exist as a nation, the principles upon which our Government is founded must not be destroyed, which they surely would be if we became nation of infidels … [and] we set the Bible aside as being untrue and put evolution in its place" (Dayton's amazing trial, 1925). Butler's bill became famous as the Butler Act, of which Butler later said, "I wrote it out just like I wanted it, and that's the way the law stands now, just the way I first wrote it" (Linder, 2004). Butler, who believed that there is no conflict between science

The Butler Act was drafted by Tennessee legislator John Washington Butler (right), who is shown here talking with William Jennings Bryan in the law offices of Sue and Herbert Hicks in Dayton (RCHGS).

and the Bible, introduced his bill (Tennessee House Bill 185) on January 21, 1925. He later admitted "I didn't know anything about evolution when I introduced it. I'd read in the papers that boys and girls were coming home from schools and telling their fathers and mothers that the Bible was all nonsense." Butler was convinced that his bill would protect children from infidelity.

One week later, the Tennessee House of Representatives—without public hearings or even the pretense of debate—passed Butler's bill by a vote of 71 to 5. Some legislators voted for it because they supported it, while others voted for it because they feared the consequences of opposing it. A Tennessee Senate committee then rejected Butler's bill, but the bill's supporters soon began a campaign to revive the bill. In February 1925, famed evangelist Billy Sunday led a massive, 10-day revival in Memphis in which he praised the Tennessee House for resisting "that God forsaken gang of evolutionary cut-throats" and condemned education as being "chained to the Devil's throne" (Linder, 2004). Sunday's revival changed several senators' minds, and on March 12 the reconvened Tennessee Senate passed Butler's bill by a vote of 24 to 6. Governor Austin Peay believed the law was unnecessary and would never be enforced, but he needed legislative help with his progressive educational program. So, on March 21, 1925, Peay signed Butler's bill into law, reminding reporters that "the people must have the right to regulate what is taught in their schools."

In July, Butler—after being praised by his home county court for his "firm and courageous stand" on the evolution issue—came to Dayton to cover the Scopes Trial for a press syndicate. When the trial began, Butler told reporters, "Ninety-nine people out of a hundred in my district thought just as I did.... There isn't a one in the whole district that thinks evolution—of man, that is—can be the way scientists tell it" (Linder, 2004). Newspaper accounts of the trial noted Butler's friendliness and kindness, as well as his surprise when Scopes' defenders mentioned English translations of the Bible other than the King James version with which he was familiar. Like most others, Butler did not foresee the worldwide interest in the Scopes Trial, adding, "I never had any idea my bill would make such a fuss. I just thought that it would become a law, and that everybody would abide by it.... I naturally thought we wouldn't hear any more about evolution in Tennessee." In Dayton, Butler insisted that he wanted his sons to read about evolution, not in school, but at home (Allen, 1925; Author of the law, 1925). After interviewing Butler, Kansas publisher Marcet Haldeman-Julius claimed, "It is impossible to conceive of a more glaring, startling plain example of the danger of sheer ignorance. [Butler] is a man, innately good, doing all but irreparable harm. A man who stands for progress in farming and in his community generally, attempting without malice to stop thought" (McRae, 2010).

After the first week of the trial, Butler told reporters that he assumed that all "right-thinking" people believed in the Bible and "the scientists have not shaken my faith." However, Butler surprised many people when he told reporters that he opposed Judge Raulston's decision barring scientists from testifying, claiming that "the judge ought to give 'em a chance to tell what evolution is. 'Course we got 'em licked anyhow, but I believe in being fair and square and American. Besides, I'd like to know what evolution is myself." When Scopes was convicted, Butler urged Judge Raulston to impose a fine of $500 for "the gravity of the offense," and his last dispatch from Dayton emphasized that he was not afraid of hearing scientists' testimony because "the Truth is mighty and will prevail ... [the evolution question] is the controversy of the age" and the "Dayton trial is the beginning of a great battle between infidelity and Christianity." After adding that "if I had it to do all over again, I'd have introduced it two years earlier—in my first term in the legislature," Butler went home, quit politics, and resumed farming (Allen, 1925).

Two years after the trial, Butler explained that he drafted his famous law after noticing that children "returned home [from school] without any respect for the faith of their Christian fathers and mothers, thereby breaking the hearts of their parents, by telling them that the Bible is not true, that no one believes the Bible except ignorant and uneducated people; that the story of creation, as taught in the Bible is only a fairy tale; [and]

that man came here not as the book of Genesis says, but by a process of evolution" (Shapiro, 2013). Butler got "a good bit of mail from all parts of the world" about the trial for the rest of his life (Antievolution law author dies in Macon, 1952). He died in his home in Lafayette, Tennessee, on September 24, 1952, and is buried in Lafayette's Butler Cemetery.

Repealing the Butler Act

The first major attempt to repeal the Butler Act came in 1929. When Tennessee state representative C.R. Morse's repeal bill was tabled, the *Nashville Tennessean* reported that "a thundering majority" of representatives had ensured "that Tennessee will remain a 'Fundamental' state under the standards of William Jennings Bryan" (Hatcher, 1929). Other attempts to repeal the law soon followed:

> 1931: Representative Robert Murray of Carroll County announced that students have a "natural right" to learn about evolution, but his bill to repeal the Butler Act was defeated in the House by a vote of 58 to 14.
>
> 1935: After the House of Representatives passed a resolution urging schools to subordinate the teaching of science to religion, it began considering Representative Cecil Anderson's bill to repeal the Butler Act. Anderson was opposed representative Sue Hicks, who had prosecuted Scopes in Dayton a decade earlier. To Hicks, evolution remained an unproven theory that would undermine students' religious faith. After representative Robert Brown proclaimed, "If [the Butler Act] was good enough for William Jennings Bryan, [it's] good enough for me," the House voted 67 to 20 to table Anderson's bill. Soon thereafter, John Butler—the bill's author—confessed, "If anyone has ever been ashamed of the [Butler Act] except infidels, I do not know who they are" (Bailey, 1953).
>
> 1939: The Tennessee Senate rejected another attempt to repeal the Butler Act, this one including an admission by one senator that some of his constituents also "believe in and practice witchcraft, but that is their business. I want to allow scientific men the same privilege." When told that Tennessee had been ridiculed as "The Monkey State" for its antievolution law, another senator reminded his colleagues, "When Christ was here he was laughed at" (House upholds state ban, 1939). By "here," the senator presumably meant on Earth, not in the Tennessee legislature.
>
> 1942: Resistance to another attempt to repeal the Butler Act is led by Bryan College, whose president, Judson Rudd, sent copies of Bryan's undelivered speech from 1925 to every legislator.

By the time other attempts to repeal the Butler Act failed in 1952 and 1959, the Butler Act had become a largely symbolic, unenforced, law, just as Governor Peay had predicted in 1925.

Things changed in 1967 when, on April 13—as a monkey looked on from the House gallery—the Tennessee House of Representatives passed a repeal bill by a vote of 59 to 30. The Senate's vote of 16 to 16 then stalled the legislation. However, near the same time in rural Campbell County, first-year science teacher Gary Scott was fired (but not arrested) for violating the Butler Act. Scott sued for reinstatement and sought, with the National Education Association and the ACLU, a permanent injunction against enforcement of the law. Scott was joined in the lawsuit by two of his students and 59 professors, associates, and graduate students in Tennessee's schools. State Senator Tom Harris, a Democrat from Chattanooga, tried to save Butler's bill by introducing an amendment that would have prohibited the teaching of evolution as "fact," but his amendment failed.

When Scott's dilemma made news across the country, John Scopes told reporters, "Those who are so violently raising Cain about this are using the power of the state to make everybody fundamentalists. They had better think about protecting liberties instead of trying to force their ideas on someone else." Scopes then added, "If teaching evolution is the only grounds for firing the fellow, I'd say it's a ridiculous thing to do. It shocks me that in this age ... that would be an excuse for firing anybody" (Porter, 1967; New monkey case evolves, 1967). The Campbell County School Board, facing bad publicity and an expensive legal fight, reinstated Scott, with back pay.

As one legislator admitted that the Butler Act "has brought us nothing but trouble," the Senate revoted 20 to 13 to repeal the Butler Act. Scott dropped his lawsuit when, on May 13, 1967, a bill (House Bill 48) repealing the Butler Act passed both houses of the Tennessee legislature. Four days later, Governor Buford Ellington—without comment—signed the repeal bill into law. For the first time in more than four decades, it was again lawful to teach evolution in Tennessee's public schools. John Scopes was the only teacher ever prosecuted for violating the Butler Act.[7]

Sources

Allen, Leslie H. 1925. *Bryan and Darrow at Dayton*. New York: Arthur Lee and Company, 1–2.

Antievolution law author dies in Macon. 1952. *Nashville Banner*, September 25, 10.

Author of the law surprised at fuss. 1925. *New York Times*, July 18, 2.

Bailey, Kenneth Kyle. 1953. The Anti-Evolution Crusade of the Nineteen-Twenties. Ph.D. Dissertation, Vanderbilt University, Nashville, Tennessee, 53.

Dayton's amazing trial. 1925. *Literary Digest*, 86 (4) (July 25), 5–7.

Hatcher, J. 1929. Bill to repeal evolution bill is killed in house. *Nashville Tennessean*, January 17, 1, 6.

House upholds state ban on teaching of evolution. 1939. *Nashville Banner*, February 16, 1.

Linder, Doug. 2004. John Washington Butler. Accessed July 29, 2020, at http://law2.umkc.edu/faculty/projects/ftrials/scopes/SCO_BUT.htm.

McRae, Donald. 2010. *The Old Devil. Clarence Darrow: The World's Greatest Trial Lawyer*. New York: Simon & Schuster, 391.

New monkey case evolves in Tennessee. 1967. *Asbury Park Sunday Press*, April 16, 6.

Porter, Michael. 1967. New firing no surprise to Scopes. *Akron Beacon Journal*, April 18, 1.

Shapiro, Adam R. 1925. *Trying Biology: The Scopes Trial, Textbooks, and the Antievolution Movement in American Schools*. Chicago: University of Chicago Press, 89.

Byrd, Howard Gale 1893–1973

Howard Gale Byrd was born on December 12, 1893. At the time of the Scopes Trial, Byrd had lived for three years in Dayton, where he was a modernist Methodist preacher. Trial instigator George Rappleyea taught Sunday School at Byrd's Five Points Methodist Church just outside of Dayton. When Rappleyea was shunned after members there learned of his views about evolution, he joined Byrd's other church, Dayton's First Methodist Episcopal Church (North) in town. Byrd, who with Rappleyea were "the only evolutionists in town," led the prayer at Scopes' hearing on May 25 (Ousted pastor, 1925). Byrd had claimed that "Christ had nothing to say about evolution; why can't we follow the example of the Master?" Nevertheless, Byrd—who told visitors that in Dayton "an evolutionist is an infidel"—often preached about evolution. When he did, however, he replaced the word "evolution" with "progress" and "development," and "the congregation regarded it as orthodox" (Dayton pastor quits, 1925; de Camp, 1968).

Byrd believed the Bible is divinely inspired, but he rejected it being "written in Heaven, printed on India paper, sewed with silk, bound with calfskin, and tossed out the

Howard Byrd (left) was a modernist preacher at Dayton's First Methodist Episcopal Church (North) who accepted evolution. When Byrd invited New York Unitarian preacher Charles Francis Potter (right) to preach about evolution in his church on July 12, Byrd's congregation rebelled and Byrd quit his job. This photograph shows Byrd, his children (John, at left, and Lillian, in Potter's arms), and Potter talking in front of the church parsonage after Byrd quit (Smithsonian Libraries and Archives, SIA RU007091).

window." Instead, Byrd urged people to be less concerned about "whence we come and whither we go" and more attentive to the "fact that we are here and here for a purpose" (Evolution trial, 1925). Rappleyea later claimed that Byrd's ideas had inspired him to initiate the Scopes Trial (Allem, 1959; Rappleyea says Byrd, 1925).

In July 1925, New York City's Unitarian preacher Charles Potter came to Dayton to help the defense team, and Byrd invited him to speak about evolution at his church. (Byrd extended a similar invitation to Foster Taylor, a fundamentalist pastor of Miami's Cocoanut Grove Presbyterian Church, of which Bryan was a member; see Evolution upsets pastorate, 1925.) When Byrd's congregation learned of Potter's upcoming sermon, it was outraged and threatened to "wreck the church" if Potter spoke there. Potter responded by posting a sign at Robinson's Drug Store announcing that his sermon had been cancelled "owing to the threat of the congregation." Byrd then announced that his congregation's "state of unfairness requires another pastor," adding, "I have quit [as pastor]. I have not resigned; I have quit."[8] Byrd's congregation did not try to change his mind (Dayton pastor quits, 1925).

When Bryan died in Dayton five days after the trial, Byrd told reporters, "A great man has fallen.... Mr. Bryan has done great good in the world." Near the same time, Byrd "was knocked out ... in a fist fight with a fundamentalist" (Acclaim from millions pours in, 1925; Modernist pastor at Dayton floored by fundamentalist, 1925).

After pointing out to reporters, "My name is spelled with a 'y' ... Evolution has carried me beyond the feathered species," Byrd and his family left for Alabama, where he hoped to work as a cabinet-maker until he could find a more liberal church. Byrd later enrolled in DePauw University and pastored churches in Alabama. He died on October 12, 1973, in Giles, Virginia, and is buried in Mt. Hope Church Cemetery in Mount Hope, Virginia.

Sources

Acclaim from millions pours in on Dayton for noted dead. *Selma (AL) Times*, July 28, 1–7.

Allem, Warren. 1959. *Backgrounds on the Scopes Trial at Dayton, Tennessee.* Master's Thesis, University of Tennessee, Knoxville, Tennessee.

Dayton pastor quits when congregation bars modernist. 1925. *Des Moines Register*, July 13, 1–2.

de Camp, L. Sprague. 1968. *The Great Monkey Trial.* New York: Doubleday, 69.

Evolution trial will begin soon. 1925. *Raleigh (NC) News and Observer*, July 3, 2.

Evolution upsets pastorate of Dayton Methodist Church and pastor abruptly resigns. 1925. *Charlotte (NC) Observer*, July 13, 1.

Modernist pastor at Dayton floored by fundamentalist. 1925. *Atlanta Constitution*, July 26, 1.

Rappleyea says Byrd responsible for case. 1925. *Chattanooga Times*, July 1, 1.

A Civic Biology
1914

A Civic Biology: Presented in Problems (1914) was a biology textbook written by George William Hunter. As its title suggests, Hunter's book framed biology with discussions of social issues such as food safety, alcohol, microbial disease, sanitation, and sexual health. *A Civic Biology* was not the first textbook to include the word "biology" in its title,[9] but it introduced a new way of teaching biology and a new generation of textbooks, for it was the first textbook to integrate botany, zoology, and human physiology into a human-centered curriculum.[10] Hunter's new approach was successful, and by the end of the decade his textbook was a bestseller. Other textbooks—including those used in colleges—soon adopted the same approach (Shapiro, 2008).

Tennessee approved *A Civic Biology* for use in its public schools in 1919, and the book was soon used in 90 percent of Tennessee's schools. Tennessee's contract to use Hunter's book expired in 1924, but —to save money—Tennessee continued to use the book in the 1924–25 school year. Tennessee governor Austin Peay was not concerned about Hunter's treatment of evolution, noting, "I can find nothing of consequence in the books now being taught in our schools with which [the Butler Act] will interfere in the slightest manner" (Peay, 1925). However, Scopes' prosecutor William Jennings Bryan disagreed, telling his friend W.B. Marr, "If you have not read [Hunter's book] I suggest you get it. It certainly gives all the ammunition we need" for Scopes' trial (Bryan to Marr, 1925). The Butler Act, which Scopes was accused of violating, did not include the word *evolution*, but Hunter's book did.

A Civic Biology, like many other biology textbooks of the era, used evolutionary biology to justify social and political agendas. For example, after discussing how selection can improve the breeding of domestic plants and animals, Hunter stressed the importance of inheritance to human society because, as he noted in a section of his book titled "Blood Tells," "brilliant men and women" are the products of "good inheritance from their ancestors." Hunter's social agenda also promoted White supremacy; in his discussion of "The Races of Man," Hunter proclaimed Caucasians "the highest" race of all.

To improve society, Hunter endorsed using selection and eugenics. Eugenics was not mentioned during Scopes' trial, but Hunter's book declared it "criminal to hand down to posterity" traits such as epilepsy, adding that "hundreds of families ... [are] spreading disease, immorality, and crime to all parts of this country ... [and] have become parasitic on society." From this flowed Hunter's "Remedy": "If such people were lower animals, we would probably kill them off to prevent them from spreading. Humanity will not allow this, but we do have the remedy of separating the sexes in asylums or other places and in various ways preventing intermarriage and the possibilities of perpetuating such a low and degenerate race. Remedies of this sort have been tried successfully in Europe and are now meeting with some success in this country."

On the Thursday of the trial, Scopes' defender Dudley Field Malone claimed that the public's interest in the trial was not "whether John Scopes taught a couple of paragraphs out of [Hunter's] book." The prosecution felt differently, for its case focused less on Scopes than on Hunter's textbook. Three of the prosecution's witnesses testified about Hunter's textbook:

> Walter White, the school superintendent and the prosecution's first witness, testified that Scopes "had taught [from *A Civic Biology*] ... [and] said that he could not teach that book without teaching evolution."
>
> Harry Shelton, one of Scopes's former students, then testified that Scopes had taught from *A Civic Biology*.
>
> F.E. Robinson, who sold Hunter's textbook at his drugstore, also testified that Scopes had admitted to teaching from *A Civic Biology*.

No one testified that Scopes had taught that humans evolved from a nonhuman ancestor. The witnesses simply said that he taught from *A Civic Biology*. However,

> nowhere did Hunter's book explicitly claim that humans descended from, or shared ancestors with, non-human species, nor did it mention any such human ancestors (e.g.,

> Cro-Magnon, Java Man, Neanderthals, or Java Man, as were mentioned in competing textbooks).
>
> Hunter classified humans as mammals primarily because of physiological and anatomical similarities, not common descent, and he did not explicitly claim that classification implies common ancestry.
>
> Although Hunter's "Evolutionary Tree" (p. 194) fit scientific conventions and had no special place for humans, Bryan complained that it grouped humans "with 3,499 other mammals.... No circle is reserved for man alone.... What shall we say of the intelligence, not to say religion, of those who ... put man with an immortal soul in the same circle with the wolf, the hyena, and the skunk?" [Bryan and Bryan, 1925].

To the prosecution, Scopes was guilty because he taught from *A Civic Biology*, the state-approved textbook. That book included evolution, so if Scopes taught from that book, he had broken the law. Although Scopes did not testify at his trial, he later claimed, "I didn't know, technically, whether I had violated the law or not." The biology exam that Scopes gave on April 28 included questions about eugenics and the major groups of organisms, but none specifically about evolution (Moore and McComas, 2016). However, several of Scopes' students, including those in his biology, general science, and physics classes that did not use *A Civic Biology*, were certain that he *had* taught evolution in their classes (*Of Monkeys and Men*, 2021). During the trial, when Darrow urged Judge Raulston to replace the "Read Your Bible" signs with ones saying, "Read your evolution" or "Hunter's Biology," Bryan responded, "We are both not swearing by Hunter's Biology. We are swearing by the Bible." In Louisiana, after the state Baptist Convention asked the state to eliminate the teaching of evolution in public schools, the state Superintendent of Education announced that he had asked the publishers of (and Louisiana's teachers who were using) *A Civic Biology* to eliminate the offending pages (i.e., pp. 191–196; Shipley, 1927).

After the Scopes Trial, Hunter's textbook—tainted by its association with the trial—was dropped by the Tennessee Textbook Commission. Joining the list of acceptable books was Peabody and Hunt's *Biology and Human Welfare* (1924), which did not discuss evolution, but praised Charles Darwin as a "great naturalist ... to whom the world owes a great part of its modern progress in biology." The next year, Hunter named his revised book *New Civic Biology*. Although Hunter demanded that his revised book include evolution, the word "evolution" disappeared from the text, as did several diagrams and references related to evolutionary concepts. Equivocation replaced certainty as every statement about Darwin and evolution was qualified with words such as "suggested" and "believed"; Darwin's "wonderful discovery of the doctrine of evolution" was changed to "his interpretation of the way in which all life changes." Mentions of natural selection (p. 383) and homology (pp. 237–238, described as "likeness in function regardless of origin") remained, as did a mention of Darwin in a section titled "Great Names in Biology" (pp. 411–412) in which "evolution" was replaced with "heredity and development."

In Hunter's revised book, Caucasians were no longer described as "the highest type of all," "lower in mentality" was changed to "lower in civilization," and the section titled "Evolution of Man" was renamed "Development of Man." Hunter expanded his discussion of eugenics (Chapter 32), but eliminated the phrase "if such people were lower animals...." Hunter continued to stress that a person's success depended on heredity, and the teacher's manual that accompanied his textbook claimed that the information about eugenics was "of more lasting value than any other part of the course" (Hunter, 1927). Hunter's revised book remained on Tennessee's state-approved list of biology textbooks until 1933.

The removal of evolution from *A Civic Biology* and other biology textbooks after the Scopes Trial caused researchers to challenge the popular claim that the defense had won the case. As Grabiner and Miller (1974) concluded, "as far as teaching biology in the high schools was concerned … [the evolutionists] had lost. Not only did they lose, but they did not even know they had lost."

Sources

Bryan, William Jennings, to W.B. Marrs [Marr], June 11, 1925. Copy in Ewing C. Baskett to Clarence Darrow, January 6, 1934, Box 2, Clarence Darrow Papers, Manuscripts Division, Library of Congress, Washington, D.C.

Bryan, William Jennings, and Mary Baird Bryan. 1925. *The Memoirs of William Jennings Bryan*. Philadelphia: United Publishers of America, 535.

Grabiner, Judith V., and Peter D. Miller. 1974. Effects of the Scopes Trial: Was it a victory for evolutionists? *Science*, 185, 832–837.

Hunter, George William. 1914. *A Civic Biology: Presented in Problems*. New York: American Book Co.

______. 1926. *New Civic Biology: Presented in Problems*. New York: American Book Co.

______. 1927. *Teachers' Manual to Accompany* a New Civic Biology. New York: American Book Co.

Hunter, George W., to George W. Benton. December 3, 1925. Syracuse: American Book Company Archives, Syracuse University Libraries, Syracuse, NY.

Moore, Randy, and William McComas. 2016. *Images of America: The Scopes Monkey Trial*. Mt. Pleasant, SC: Arcadia Press, 32.

Of Monkeys and Men: Public and Private Views of the Scopes Trial. 2021. Special Collections, University of Tennessee, Knoxville, Tennessee.

Peabody, James Edward, and Arthur Ellsworth Hunt. 1924. *Biology and Human Welfare*. New York: Macmillan, 6, 534, 548–549.

Peay, Austin M. 1925. Austin M. Peay to Tennessee House, printed in *Journal of Tennessee House*. Nashville: State of Tennessee, 743–745.

Shapiro, Adam R. 2008. *Civic Biology* and the origin of the school antievolution movement. *Journal of the History of Biology*, 41 (3), 430.

______. 2013. *Trying Biology: The Scopes Trial, Textbooks, and the Antievolution Movement in American Schools*. Chicago: University of Chicago Press.

Shipley, Maynard. 1927. *The War on Modern Science: A Short History of the Fundamentalists' Attacks on Evolution and Modernism*. New York: Alfred A. Knopf, 82–85.

Colby, Bainbridge
1869-1950

Bainbridge Colby was born into a wealthy family on December 22, 1869, in St. Louis, Missouri. He graduated from Williams College in 1890, after which he attended New York Law School in 1892. He was then elected to the New York State Assembly and in 1912 helped form the Progressive "Bull Moose" Party. In 1916, Colby campaigned for the reelection of Woodrow Wilson, and in 1920 he became Wilson's third Secretary of State. After resigning the next year (i.e., at the end of Wilson's presidency), Colby resumed his private law practice in Washington, D.C.

Unbeknownst to the ACLU, George Rappleyea asked Colby to join the team defending John Scopes. Colby initially agreed and, after a visit to Dayton, denounced the trial's "holiday atmosphere" while claiming that the trial was the foundation "of our security of happiness and American citizenship" (Scopes Trial food, 1925). However, Colby was uncomfortable in Tennessee; after admitting to a reporter, "This is no place for me," he resigned from the defense team two days before the trial started. Colby was replaced by Darrow's law partner, William O. Thompson, who played a minor role in the trial.

Colby, the personal attorney for Mark Twain, died on April 11, 1950, in Bemus Point, New York. He is buried there in Bemus Point Cemetery.

Source

Scopes Trial food for thought, Colby. 1925. *Knoxville Journal*, June 25, 1.

Dalhart, Vernon
1883–1948

Marion Try Slaughter II, better known by his stage name (Vernon Dalhart), was an America songwriter and country music singer. He was born in Jefferson, Texas, and later based his stage name on two Texas towns, Vernon and Dalhart. In 1910, Dalhart moved to New York City, where he sang tenor in several operas. American inventor Thomas Edison, the owner of Edison Records, signed Dalhart to his first record contract (Palmer, 2004).

On July 10, 1925—that is, on the first day of the Scopes Trial—Dalhart was in New York City recording Carson Robison's "The John T. Scopes Trial (The Old Religion's Better After All)":

> Oh the folks in Tennessee are as faithful as can be,
> And they know the bible teaches what is right.
> They believe in God above and His great undying love,
> And they know they are protected by His might.
> Then to Dayton came a man with his new ideas so grand,
> Who said we came from monkeys long ago,
> But in teaching his belief, Mister Scopes found only grief.
> For they would not let their old religion go.
>
> CHORUS: You may find a new belief, it will only bring you grief,
> For a house that's built on sand is sure to fall.
> And wherever you may turn, there's a lesson to be learned,
> That the old religion's better after all.
>
> Then the folks throughout the land saw his house was built on sand
> And they said, "We will not listen any more."
> So they told him he was wrong, and it was not very long
> That he found that he was barred from every door.
> Oh, you must not doubt the word that was written by the Lord
> For if you do your house will surely fall
> And Mr. Scopes will learn that, wherever he may turn
> That the old religion's better after all.

Dalhart, who was accompanied by guitar and fiddle, recorded three takes of the song. His record was released in early August; it sold for 75¢ and, riding the trial's lingering publicity, eventually sold more than 400,000 copies. On August 10, Dalhart returned to the studio to record "William Jennings Bryan's Last Fight," which was released later that month (Palmer, 2004). Dalhart rerecorded these songs for several other labels, including Cameo (No. 792) and Gennett/Silvertone (No. 3134). "The John T. Scopes Trial" was also recorded by other artists, including blind minstrel Charles Oaks and Charles Mabell. Other songs related to the Scopes Trial that were popular near this time included "The Death of William Jennings Bryan," "The Bible's True," "There Ain't No Bugs on Me," "I'm No Kin to the Monkey," "The Monkey Case," and "You Can't Make a Monkey Out of Me" (Wilhoit, 2000).

Dalhart's career slumped with the Depression, and by 1942 he was working as a clerk at Barnum's Hotel (now Barnum House) in Bridgeport, Connecticut. Dalhart suffered a heart attack in January 1948, and died after a second attack on September 14, 1948, in Bridgeport. He is buried in Bridgeport's Mountain Grove Cemetery under his given name, not his stage name. Although largely forgotten by the public, Dalhart's obituary appeared in virtually every major newspaper in the United States.

Dalhart, who used more than 100 aliases and recorded for several different labels, sang on more than 5,000 records. His recording of "Wreck of the Old 97" was the first country song to sell more than a million copies.[11] Dalhart, whose recordings have sold over 10 million copies, was posthumously elected to the Nashville Songwriters Hall of Fame in 1970, and the Country Music Hall of Fame in 1981. Along Jefferson's West Austin Street, a historical marker commemorates Dalhart and how his music "led to the rise of country music as an American art form."

Sources

Palmer, Jack. 2004. *Vernon Dalhart: First Star of Country Music.* Littleton, CO: Mainspring Press.

Wilhoit, Melvin R. 2000. "You Can't Make a Monkey Out of Me," or forgotten voices from the sticks: Music of the Scopes Monkey Trial. Pp. 54–69 in Cornelius, Richard M., and Tom Davis. *Impact: The Scopes Trial, William Jennings Bryan, and Issues That Keep Revolving.* Dayton, TN: Bryan College.

Darrow, Clarence Seward 1857–1938

Clarence Darrow was born on April 18, 1857, in a small village outside of Kinsman, Ohio. Amirus, Darrow's politically motivated father, was a furniture maker and a free-thinking abolitionist, and Emily, his mother, was an early supporter of women's rights. After attending Allegheny College and the University of Michigan law school, Darrow taught for three years and studied law. He was admitted to the Ohio bar in 1878 and began practicing law. After moving to Chicago in 1887 and working there for the city, he began representing the Chicago North-Western Railway Company, but in 1894 he quit that job to represent Eugene Debs (leader of the American Railway Union) against federal charges involving the Pullman Strike. Debs, who Darrow admitted was "one of my heroes," went to jail for six months.

Darrow had been interested in political questions since he was a child, and in 1895 he ran for Congress as a Democrat, despite admitting to having "few illusions to what [that] meant." He campaigned little in his district and lost. The next year, Darrow attended the Democratic National Convention in Chicago and heard Bryan deliver his famous "Cross of Gold" speech that secured his party's nomination for president. Decades later, Darrow recalled that he had "never listened to [a speech] that affected and moved an audience as [Bryan] did," adding that it was followed by "the greatest ovation I had ever witnessed" (Darrow, 1932). Although Darrow questioned politicians' appeal to "the people—with all its ignorance, its prejudice, its selfishness, and, worst of all, its insincerity as to either men or principles" (Darrow, 1932), he then campaigned for Bryan, but Bryan lost.

From 1906 to 1908, Darrow helped get William "Big Bill" Haywood and other mining executives acquitted of conspiring to commit murder. In 1911, Darrow defended American Federation of Labor leaders James and John McNamara, who were charged with killing 21 workers when they dynamited the headquarters of the fiercely anti-labor *Los Angeles Times*. The McNamara brothers went to prison, and on January 30, 1912, Darrow was indicted on two counts of bribing a juror in the trial. Evidence suggested that Darrow was guilty, but he pleaded that his conviction would harm "the weak and the poor and the helpless throughout the world." Darrow was acquitted on one count, and the other count produced a hung jury (Cowan, 1993; McRae, 2009). The judge agreed to not retry Darrow if Darrow promised to never again practice law in California.

In 1924, Darrow—a fierce critic of the death penalty—defended Nathan Leopold, Jr., and Richard Loeb, two privileged teenagers from rich families charged with kidnapping and "thrill-killing" teenager Bobby Franks. Leopold and Loeb considered the murder "an intellectual exercise" that "was just as easy to justify … [as] an entomologist killing a beetle on a pin" (McRae, 2010). The murder outraged the public, but Darrow's closing argument—parts of which continue to be invoked by lawyers representing defendants charged with murder—convinced the judge to send his clients to prison rather than the gallows. (Darrow's argument also inspired the 1929 play, *Rope*; see Dershowitz in Darrow, 1932.) Just as the Leopold-Loeb trial had given Darrow a prominent platform from which to condemn capital punishment, Darrow would use the Scopes Trial to condemn what he felt was religion's oppression of education and free thought.

Darrow made no attempt to hide his agnosticism. He knew that the existence of God could not be disproven, and claimed that he was agnostic because "I am not afraid to think. I am not afraid of any God in the universe who would send me or any other man or woman to hell. If there were such a being, he would not be a God; he would be a devil" (Clarence Darrow, noted lawyer, succumbs at his home, 1938). Despite his agnosticism, Darrow admitted that he had "never seen a time he wouldn't welcome Jesus Christ on a jury" (Adamson, 1975). Darrow, who got rich by defending wealthy, privileged clients, also often fought for underdogs and against public opinion, which he considered "the greatest enemy that has ever confronted man" (Darrow, 1932).

By the early 1920s, Darrow detested Bryan's shift to fundamentalism, and was determined to expose Bryan's views. After publicly attacking several universities for teaching evolution, Bryan wrote to the *Chicago Tribune*, "It is improper and subversive … to teach atheism or agnosticism or true Darwinism or any hypothesis that links man in a blood relationship with any other form of life" (McRae, 2009). In response, Darrow composed a letter that was printed on the front page of the July 4, 1923, issue of the newspaper. Darrow's letter, which criticized Bryan's attacks on evolution and his efforts to shut out the teaching of science, included "a few questions to Mr. Bryan and the Fundamentalists [which], if answered, might serve the interests of reaching the truth—all of this assuming that the truth is desirable." Darrow's 55 questions addressed biblical literalism; for example, was the earth made in six literal days? How did Noah gather animals from all parts of the world to the Ark? Was Eve literally made from Adam's rib? How could species found only in Australia or other distant places get to the Ark, and why did they not stop on the way? Can someone be a Christian without believing in the literal truth of the Bible? Bryan, who that night was addressing a crowd of almost 10,000 people at Billy Sunday's tabernacle in Winona Lake, Indiana, ignored Darrow's questions, explaining, "I decline to turn aside to enter into controversy with those who reject the Bible as Mr. Darrow does" (Stone, 1943). Bryan, who believed that Darrow was "the only man who looks down on me for believing in God," added, "I am not worried about an atheist who admits he is one. The man who denies the existence of God is not likely to have much influence. My controversy is not with atheists like Mr. Darrow, but with those who claim to be Christians and who substitute the guesses of evolution for the word of God.... The unproven hypothesis of evolution is the root cause of nearly all the dissension in the church" (McRae, 2009).

In 1925, when Bryan accepted the World's Christian Fundamentals Association's invitation to help prosecute John Scopes, Darrow admitted, "At once I wanted to go" to Dayton to defend Scopes, adding, "My object, and my only object, was to focus the attention of the country of the program of Mr. Bryan and the other fundamentalists in America" (Darrow, 1932). Although the ACLU feared that Darrow's "headline chasing" would detract from the constitutional issues that it wanted to test, Scopes insisted that Darrow defend him (Scopes and Presley, 1967).

In June 1925, Darrow—described as "a stooped man in baggy dark clothes"—went to Tennessee and warned graduates of John Randolph Neal's law school, "The country has fallen upon evil times. It seems that every organization has some law it is endorsing to force upon the people," adding that he hoped, "Some day we may have a movement to let people alone." Darrow's subsequent lecture to the Tennessee bar association was cancelled when members began arguing about Scopes' upcoming trial (Evolution stalks state bar meeting, 1925). Darrow then visited Dayton, where his first stop was Robinson's Drug Store so he could "see where it started" (Darrow will meet counsel in Scopes case, 1925). Darrow denounced Tennessee's antievolution law as "as brazen and bold an attempt to destroy learning as was made in the Middle Ages," Darrow claimed that Scopes "was indicted for the crime of teaching the truth" and sarcastically wondered why "the Dayton school board did not adopt the first and second chapters of Genesis as a modern textbook of biology" (Darrow, 1932). In Dayton, before the trial even started, there were promises that Darrow would "make one of the great addresses of the trial" (Kinsley, 1925a).

Darrow soon realized that the "proceedings [in Dayton] bore little semblance to a court case" and "seemed to me a joke" promoted by "no end of publicity." When the defense tried (unsuccessfully) to move the trial to federal courts, Darrow told reporters, "This is an increasingly serious matter.... The Dayton trial has been treated with levity and it is not a matter of joking.... Many of the public do not appreciate the importance of this prosecution" (Scopes attorneys seek appeal, 1925).

Darrow was surprised when Judge Raulston opened each day's proceedings with a prayer, for during his 40 years as a lawyer he "had never before heard God called in to referee a court trial." When Darrow objected to the prayers, "the people assembled looked as though a thunderbolt had stunned them" (Darrow, 1932). Darrow, who believed that "from any point of view, the [Butler Act] was silly and senseless," opposed fundamentalism because he hoped to prevent "bigots and ignoramuses from controlling the education at the United States" (Darrow, 1932). Darrow blamed Bryan for the law, which Darrow claimed made "the Bible the yard stick to measure every man's intellect, to measure every man's intelligence and to measure every man's learning." In his most famous speech in Dayton's courtroom, Darrow proclaimed, "If today you can take a thing like evolution and make it a crime to teach it in the public school, tomorrow you can make it a crime to teach it in the private schools.... [You may then] ban books and the newspapers. Soon you may set Catholic against Protestant, and Protestant against Protestant, and try to foist your own religion upon the minds of men. If you can do one you can do the other. Ignorance and fanaticism is ever busy and needs feeding.... Today it is the public school teachers, tomorrow the private. The next day the preachers and the lecturers, the magazines, the books, the newspapers. After a while, Your Honor, it is the setting of man against man and creed against creed until with flying banners and beating drums we are marching backward to the glorious ages of the sixteenth century when bigots lighted fagots to burn the men who dared to bring any intelligence and enlightenment and culture to the human mind." *The Chicago Tribune* praised Darrow's speech as "one of the greatest speeches of his career," Scopes' prosecutor Ben McKenzie followed suit, adding that it "was the greatest speech I ever heard in my life—on any subject," and publisher Marcet Haldeman-Julius claimed that the speech left Darrow "stand[ing] a giant among mental pygmies ... one of the truly great men in America" (Kinsley, 1925b; McRae, 2010). On Monday, July 20, Judge Raulston cited Darrow's "contempt and insult" of the court on the previous Friday. Darrow later apologized and Raulston dropped the charge.

The most famous event of the Scopes Trial occurred on July 20 when the defense called Bryan to the witness stand. While questioning Bryan about Biblical history and revealed religion, Darrow acted as if Bryan, and not Scopes, was the defendant. In response to Darrow's questions, several of which came from his July 1923 letter in the *Chicago Tribune*, Bryan admitted his long-held belief that Earth was ancient. When Bryan testified that he was trying "to protect the world of God against ... the greatest atheist or agnostic in the United States," Darrow—who believed that "as to science, [Bryan's] mind was an utter blank"—replied that he was questioning Bryan's "fool ideas that no intelligent Christian on Earth believes" (Darrow, 1932). Throughout the confrontation, the large crowd was "generous with their applause to both contestants," but some were upset by what they considered Darrow's irreverence. When the confrontation ended, Darrow's colleague Arthur Garfield Hays reported, "The followers of Darrow rose up and swarmed to his side, anxious to seize the hand of their champion. Bryan stood apart, almost alone, a strained, tired expression on his face as he looked into the twilight that was closing about him" (Farrell, 2011). Darrow, who was often supported by a press hostile to Bryan, later admitted, "I truly felt sorry for Mr. Bryan" (Darrow, 1932). In an oft-quoted summary of the trial, columnist Paul Anderson told his readers that the confrontation between Darrow and Bryan—"two old men"—was "magnificent and tragic, stirring and pathetic, and above all it was pervaded by the atmosphere of grandeur which befitted a death grapple between two great ideas" (Anderson, 1925). Today, a century later, the confrontation between Darrow and Bryan remains a symbol of the societal tension between religion and science in America.

Ruby Darrow, Clarence's wife, came to Dayton for the trial, but said relatively little to reporters, aside from joking that her husband's disheveled look resulted from him "sleeping in his clothes." Later, Ruby claimed that Mary Bryan was "wheeled into the courtroom by her nurse ... [with] her dainty white gown, her beautiful countenance and graces and almost saintliness.... I am not aware that [the Bryans] had any intention to

On the afternoon of July 20 on a dais outside the courthouse, Darrow (standing, right) questioned Bryan (seated, left, wearing a bowtie) not about evolution, but instead about the literal truth of the Bible. This 90-minute confrontation remains a popular symbol of the societal tensions between religion and science in America (Watson Davis, Smithsonian Institution Archives 2007–1024).

be rude to me, but certainly they glanced the other way any time we were all near each other" (Farrell, 2011).

On Tuesday morning (July 21), Judge Raulston struck Darrow's examination of Bryan from the court record. Several newspapers claimed that Darrow had "won" the confrontation, but others disagreed; for example, in his introduction in Darrow's *The Story of My Life* (1932), Alan Dershowitz concluded in 1995 that "Darrow came off as something of an anti-religious cynic" while, for the most part, Bryan "seems to have gotten the better of Clarence Darrow in the argument over the Bible" (Darrow, 1932). Darrow asked the jury to convict Scopes, and it did. Because Darrow did not give a closing statement, neither could Bryan. Many newspapers published Bryan's undelivered closing statement (Bryan, 1925).

After the trial, Bryan issued nine questions for Darrow to answer about Christianity. Darrow's answers rejected Bryan's image of Christ, the Bible being inspired by God, and miracles. Some of Darrow's answers (e.g., "I do not pretend to know...") resembled those given by Bryan during the trial which had been ridiculed by journalists.

Although Darrow had often defended other clients *pro bono*, the Scopes Trial was the first and only time that Darrow volunteered to defend someone for free. Darrow later admitted, "I really wanted to take part in [the trial] ... [to show] the country what an ignoramus [Bryan] was, and I succeeded." Darrow, who spent about $2,000 of his own money to defend Scopes, admitted that he had "enlisted like the boys who went to France, for liberty and humanity" (Darrow, 1932; Minister here condemned Darrow's agnosticism, 1938). Darrow lamented, "Tennessee never deserved all the unfavorable notoriety that the fundamentalists brought it," and he praised the people of Dayton for being "more hospitable than we are up North." On the night the trial ended, Darrow attended a dance for journalists, attorneys, and others associated with the trial, where he was to give his lavender suspenders (and Malone was to give a silk shirt) to the journalists who were still in town. The dance was described as "a gala night" during which "the Mason and Dixon line was erased completely for the evening" (McGeehan, 1925). For the rest of his life, Darrow considered the Scopes Trial to be one of his most gratifying cases.

After the Scopes Trial, Darrow wanted to retire, claiming that he "had fought for the minority long enough" (Darrow, 1932). However, just two months after the Scopes Trial, Darrow and Arthur Garfield Hays defended Ossian Sweet, an African American who was charged with murder while defending his home from an angry White mob. Sweet's acquittal announced that Black citizens, like White citizens, could defend their homes from attackers. Darrow's last major case came four years after he had closed his law office, when he defended Thomas Massie in another racially charged murder case. Throughout his life, Darrow believed that the "one way toward any general betterment of social organization ... [is] by teaching sympathy and tolerance" (Darrow, 1932).

Several years after the Scopes Trial, Darrow returned to Dayton while on his way to speak in Florida. When he saw a new church downtown, he commented, "it doesn't look like the devil and I did very much good here" (Bryan Memorial faculty, 1930). Darrow later narrated *The Mystery of Life* (1931), a Universal Pictures movie that covered some of the material introduced by the defense team's witnesses at the Scopes Trial. In 1932, Darrow, who received letters about the Scopes Trial for the rest of his life, published his autobiography, *The Story of My Life*, in which he (like others before and since) claimed that Bryan believed that Earth was less than 6,000 years old.

When Scopes last saw Darrow in the mid–1930s, Scopes knew, "The great dynamo was slowing down; his great store of energy was running out" (Scopes and Presley, 1967). Following a long physical and mental decline, Darrow died on March 13, 1938, of heart disease in his Chicago home, surrounded by his family. In Dayton, F.E. Robinson reminisced that Darrow's death "brings to mind the epochal days of the Scopes Trial.... His sincerity and his conviction had to be admired," and Charles Jones—who had officiated Bryan's memorial service in Dayton—admitted, "Mr. Darrow was a man for whom I had great admiration." In New York City, Dudley Field Malone praised Darrow as "the most distinguished human being I've ever met," and fellow defender Arthur Garfield Hays eulogized Darrow as "the greatest of them all" (Dayton mourns death of Clarence Darrow, 1938). Darrow's ashes were scattered into Chicago's Jackson Park Lagoon, where Darrow often walked with his wife and son.

Darrow has been the subject of many books, plays, movies, and documentaries. He is also commemorated by the Clarence Darrow Memorial Bridge at Jackson Park Lagoon (just south of the Museum of Science and Industry) and a statue outside the Rhea County Courthouse in Dayton. Darrow is widely regarded as one of the greatest criminal defense attorneys in American history.

Sources

Adamson, June. 1975. Nellie Kenyon and the Scopes "Monkey Trial." *Journalism History*, 2 (3), 88–97.
Anderson, Paul Y. 1925. Science vs. fundamentalism. *St. Louis Post-Dispatch*, July 21, 1.
Bryan Memorial faculty must believe story of creation. 1930. *Chattanooga News*, March 23, 23.

Bryan, William Jennings. 1925. *Last Message*. New York: Fleming H. Revell.
Clarence Darrow, noted lawyer, succumbs at his home in Chicago. 1938. *Billings Gazette*, March 14, 8.
Cowan, Geoffrey. 1993. *The People v. Clarence Darrow: The Bribery Trial of America's Greatest Lawyer*. New York: Times Books.
Darrow, Clarence. 1932. *The Story of My Life*. Cambridge, MA: De Capo Press, vii, 52, 54, 232, 244, 247–250, 258–259, 274, 277, 301.
Darrow will meet counsel in Scopes case. 1925. *Chattanooga News*, June 22, 1.
Dayton mourns death of Clarence Darrow. 1938. *Knoxville News-Sentinel*, March 14, 3.
Evolution stalks state bar meeting. 1925. *Memphis Commercial Appeal*, June 17, 1.
Farrell, John A. 2011. *Clarence Darrow: Attorney for the Damned*. New York: Doubleday, 374–376, 395.
Kinsley, Philip. 1925a. Court may bar Scopes experts. *Ft. Worth Star-Telegram*, July 12, 1, 6.
______. 1925b. Liberty at stake if law stands, Darrow says. *Chicago Tribune*, July 14, 1.
McGeehan, W.O. 1925. Why pick on Dayton? *Harper's Magazine*, 151 (October), 623–627.
McRae, Donald. 2009. *The Great Trials of Clarence Darrow: The Landmark Cases of Leopold and Loeb, John T. Scopes, and Ossian Sweet*. New York: Harper Perennial, 150, 156–157.
______. 2010. *The Old Devil. Clarence Darrow: The World's Greatest Trial Lawyer*. New York: Simon & Schuster, 190–198–199.
Scopes attorneys seek appeal to federal jurists. 1925. *Paducah Evening Sun*, July 4, 1.
Scopes, John T., and James Presley. 1967. *Center of the Storm: Memoirs of John T. Scopes*. New York: Holt, Rinehart and Winston, 70, 231.
Stone, Irving. 1943. *Clarence Darrow for the Defense*. Garden City, NY: Doubleday, 426–427.

Darwin, Leonard
1850–1943

Leonard Darwin was born on January 15, 1850, in Down House, just outside of London. He was the eighth of ten children and the fourth son of Charles Robert (1809–1882) and Emma Wedgwood Darwin (1808–1896; Charles's first cousin). Following a military career in which he rose to the rank of Major, Leonard served in Parliament, after which he was president of the Royal Geographical Society from 1908 to 1911. From 1911 to 1928, Darwin—who opposed interracial marriage and believed Blacks were inferior to Whites—chaired the British Eugenics Society. He advocated for the sterilization of people "endowed with bad hereditary qualities" to block "a contaminated tributary to the great stream of life" and thereby avoid "an immense amount of human misery, degradation, and inefficiency ... in all future generations" (Darwin, 1923). Leonard never formally studied science, but he was a friend and confidant to fellow eugenicist Henry Fairfield Osborn, who helped the defense team before and during the Scopes Trial. Horatio Newman, a eugenicist whose testimony was read into the record of the Scopes Trial, often promoted Leonard Darwin's harsh eugenic beliefs (Newman, 1932). Leonard—who believed that eugenics was the practical application of evolution—presided at the First International Congress of Eugenics (July 24–29, 1912, in London). At the second such congress (September 25–27, 1921, in New York City), Darwin promoted "eliminating the unfit." He was convinced that only eugenics could save Western civilization.

Darwin closely followed the events leading to the Scopes Trial. On June 23, 1925, Leonard wrote a letter to John Scopes assuring him that "that which is true cannot be irreligious.... May the son of Charles Darwin send you in his own name one word of warm encouragement" (Blaney, 2011). After the trial, Darwin continued to promote eugenics. In 1926, he dedicated his 529-page *The Need for Eugenic Reform* to his father's memory, writing, "If I had not believed that he would have wished me to give such help as I could towards making his life's work of service to mankind, I should never have been led to write this book" (Darwin, 1926).

The 93-year-old Leonard Darwin—Charles and Emma's last surviving child—died on March 26, 1943, in Forest Row, East Sussex, where he is buried in Forest Row Cemetery.

Sources

Blaney, Tom. 2011. *The Chief Sea Lion's Inheritance: Eugenics and the Darwins*. Leicester, UK: Matador.
Darwin, Leonard. 1923. Sterilization in America. *The Eugenics Review*, 15 (1), 335–344.
______. 1926. *The Need for Eugenic Reform*. London: John Murray.
Newman, Horatio Hackett. 1932. *Evolution, Genetics, and Eugenics*, Third Ed. Chicago: University of Chicago Press, 441–444.

Davis, Watson
1896–1967

Watson Davis, born April 29, 1896, was a pioneer in library science who promoted science to the public. After earning a civil-engineering degree from George Washington University in 1917, Davis—the son of a high school principal—worked for the National Bureau of Standards, after which he was the science editor for the *Washington Herald*. In 1921, Davis began working for Science Service, a non-profit organization that marketed science as a commodity to newspapers, magazines, and "the reading world." The following year, he became editor of *Science News Letter*, whose circulation exceeded 45,000 readers (Davis, 1949). It was in this capacity that Davis covered the Scopes Trial. According to Davis, trial instigator George Rappleyea was "our Dayton correspondent."

In June 1925, Davis visited Dayton to interview Rappleyea and Scopes. A few weeks later, while returning to Dayton with his colleague Frank Thone to cover the trial, Davis sent telegrams—in the names of Malone and "Clarance" Darrow—inviting scientists to come to Dayton to help defend Scopes. Davis, who during the trial stayed with the scientists in "The Mansion," claimed that there was no conflict between science and religion. He also believed that "education is the real job of the defense," promising, "The world ... [would] take one of the most interesting and vital correspondence courses that the newspapers have ever offered to the public" (Davis, 1925b). When the trial began, Davis reported, "It hardly seems possible that the prosecution will succeed ... [in excluding] the scientific evidence and experts" and hoped that the trial would be "the scene of upward steps in the evolution of the human thinking mind" (Davis 1925b).

Throughout the trial, Davis—a board member of the American Eugenics Society—was prominent among the press corps, which he described as being "worthy of a prizefight or a political convention" (Davis, 1925b). Davis was decidedly anti-Bryan, and portrayed Bryan and his supporters as a dying breed of ignorant, rural bigots. Davis even convinced several of the scientists brought to Dayton to sign a letter endorsing Darrow's "ability, high purpose, integrity, moral sensitiveness, and idealism" (Darrow, 1932). To support his claim that it was "truly pitiful" to see Bryan's "gross ignorance of science," Davis incorrectly reported that Bryan claimed that humans are not mammals (Davis, 1925a). Four months after the trial, Davis also claimed, "One good thing the trial did was give us the biologist Scopes" (Scopes trial at least gave us biologists, 1925). Scopes never was, nor ever claimed to be, a biologist.

When the trial was in session, Davis shared a table with H.L. Mencken directly behind the defense team and about 15 feet from Judge Raulston. Taking a cue from Mencken, Davis created controversy early in the trial by allegedly referring to Baptists and Tennessee as "pagan" and the trial as "a new Inquisition" (Callaway, 1925). After Darrow's questioning of Bryan on July 20, Davis and Thone ridiculed Bryan's "pitiful exhibition of ignorance" while praising Darrow's "skillful examination" as "the most important evidence for the defense…. Darrow's questioning made Bryan confess that he had little real knowledge of religion and science" (Darrow, 1932). Davis's articles from Science Service reached almost 20 percent of the daily circulation of the American press. Science Service's profits from the Scopes Trial totaled $250.50 (approximately $2,250 in 2022).

After the trial, Davis and Thone—with scientists Kirtley Mather and Maynard Metcalf—helped raise money for Scopes to attend graduate school. Although Scopes' defender John Neal argued that their fundraising should wait until after Scopes' appeal, Davis and Thone moved ahead. Their goal was $5,000, but they raised only $2,550, most of which came from biologist David Starr Jordan.[12] After the trial, Davis preached in modernist churches about "Science versus Religion."

In 1933, Davis became director of Science Service, a position he held until 1966. Davis was an early advocate for the use of microfilm and became a pioneer in library and information science. He also founded the American Documentation Institute, Science Clubs of America, and the Science Talent Search to recognize and promote students' interest in science.

In 1960, Davis renewed his friendship with Scopes at Dayton's Scopes Trial Day celebrating the 35th anniversary of the trial. While in Dayton, Davis reported that the evolution controversy in Tennessee was "feebler, perhaps, but not dead" (Scopes Trial, recalled, 1960).

Davis retired in 1966 and, on June 27 of the following year, the 71-year-old Davis died of cancer in Washington, D.C. Upon hearing of his death, Rodney Wader noted, "It is possible—even probable—that no one who ever lived has worked harder, longer, or more effectively at making science intelligible to the average citizen than did Watson Davis" (Wader, 1967). Glenn Seaborg, chairperson of the Atomic Energy Commission, offered similar praise, noting, "Davis has done more for the popularization of science and the understanding of science by the general public than any other individual" (Watson Davis, editor, 1967). Several professional organizations, including the History of Science Society and Association for Information Science and Technology, sponsor awards honoring Davis.

Sources

Callaway, T.W. 1925. The Rev. T.W. Callaway glad to "eat crow." *Chattanooga Times*, July 15, 2.
Darrow, Clarence. 1932. *The Story of My Life*. New York: Charles Scribner's Sons, 261, 278.
Davis, Watson. 1925a. Man not a mammal, Bryan tells court. *Baltimore Evening Sun*, July 17, 2.
______. 1925b. School of science for Rhea County. *Chattanooga Daily Times*, July 11, 1.
______. 1949. Science Service: The popular exposition of science. *Nature*, 4146 (September 3), 399.
LaFollette, Marcel Chotkowski. 2008. *Reframing Scopes: Journalists, Scientists, and Lost Photographs from the Trial of the Century*. Lawrence: University Press of Kansas.
Scopes trial at least gave us biologists—Watson Davis. 1925. *Capital Times* (Madison, WI), November 12, 10.
Scopes Trial recalled. 1960. *Science News Letter*, July 30, 69.
Wader, Rodney. 1967. Watson Davis. *Chemical Engineering News*, 45 (29), 5.
Watson Davis, editor, at 71. 1967. *Newsday* (Melville, NY), June 28.

Dayton Coal & Iron Company
Est. 1882

On January 16, 1882, the arrival of 20 carloads of railroad rails in Dayton, Tennessee, announced the upcoming formation of Dayton Coal & Iron Company (DCIC). These rails soon comprised eight miles of tracks leading from the nearby coal mines to a blast furnace that processed the coal. In 1882, DCIC's first mine—Richland Mine, named for a nearby creek in which methane often bubbled up through the water—began DCIC's transformation of Dayton. DCIC's failure three decades later triggered an economic slump that pushed citizens to search for ways to revitalize the town.

DCIC, which was owned by textile maker Sir Titus Salt, Jr., & Sons of Saltaire, England, covered 23,845 acres of land and had the mineral rights to 3,844 more. After being officially incorporated in February 1883, the company grew rapidly and soon provided jobs for more than 2,000 workers. In 1885, just three years after the opening of Richland Mine, the 468-feet-deep Nelson Mine (named for John Nelson, who had discovered its seam of coal) opened, and the Dixon-Slope Mine—built primarily for ventilation—followed in 1891. DCIC's blast furnace could process 200 tons of pig iron per day, and an associated furnace could produce 10,000 bricks per day. Business was booming, as was Dayton's economy. (In 1890 alone, DCIC produced 210,465 tons of coal.) To support its growth, DCIC built a school, store, crating house, engine house, 200 houses for miners and their families, and a 10-bedroom, $15,000 Victorian mansion for John H. Ferguson (1840–1888), DCIC's superintendent and general manager (Broyles, 1991). By

By the early 1900s, Dayton Coal & Iron Company (DCIC, whose furnaces and smokestacks loom in the background) had made Dayton one of the post–Civil War South's leading industrial centers. In the foreground, a crowd at the Dayton train station awaits the arrival of people coming for the trial. The collapse of DCIC in 1913 put the area's economy into a tailspin that some of Dayton's businessmen hoped would be improved by a show-trial testing the Butler Act. This show-trial became the Scopes Trial (RCHGS).

1896, the success of DCIC had tripled Dayton's land-prices as the area's population grew from 1,900 to more than 6,000 people. Thanks to DCIC, Dayton had become one of the major industrial sites in the South.

Everything began to change early on the morning of Friday, December 20, 1895, when—just after a gas warning had been posted—a lamp carried by inspector Tom Hawkins triggered a spark at one of the entrances of Nelson Mine (Miners entombed, 1895). The resulting explosion killed Hawkins and 29 of the mine's other 150 workers, transforming Dayton into "a city of the dead" as it reeled from "the most appalling catastrophe the history of East Tennessee" (Broyles, 1991). Reparations paid by DCIC to grieving families averaged $192 and ranged from $10 to $500 (Chappuis, *et al.*, 2018).

Low prices and labor strikes in 1897, 1898, 1899, 1903, and 1904 also plagued DCIC's profits, and another explosion—this one in Richland Mine on May 27, 1901—killed another 35 people. Richland Mine reopened a few weeks later, but 10 months later another explosion at Nelson Mine killed 25 more miners. As DCIC's losses began to pile up, Nelson Mine closed permanently in 1908, and Richland Mine closed two years later.

DCIC's managers—drowning in debt—decided to stockpile iron, hoping that prices would increase. However, DCIC was soon overextended and, on June 13, 1913, filed for bankruptcy. Hundreds of people were left unemployed; Peter Donaldson, DCIC's president and managing director, committed suicide by chaining himself in his car and driving into the River Thames (Broyles, 1991). Subsequent attempts to resurrect DCIC, including one in 1916 led by Dayton mayor A.P. Haggard, failed. Low prices, labor strikes, and mining accidents had ended mining in Dayton, and Dayton's population shrank to just 1,800 people. Unbeknownst to DCIC, the miners had missed a massive seam of coal by only 23 feet. Had that seam been found, it might have sustained DCIC until the 1930s and, in the process, left Dayton without a reason to concoct the Scopes Trial, and left John Scopes an unknown teacher (Johnson, 2012).

By 1924, the remnants of DCIC were owned by Cumberland Coal & Iron Company, which later closed most of area's major coal-mines for good. Cumberland Coal & Iron began selling DCIC's assets, including its land, buildings, and miscellaneous equipment. Supervising this liquidation was chemical engineer George Rappleyea, who in early May 1925 would visit Robinson's Drug Store and instigate the Scopes Trial.

Most of DCIC's buildings and other structures are now gone; the company store and office complex were torn down for bricks, and the company's locomotive and furnace were cut up for scrap. The Mansion, which housed the defense team's experts during the Scopes Trial, burned to the ground in 1945. Some of DCIC's remaining sites and structures (e.g., railroad rails, bridges, and coke ovens, most overgrown by kudzu) are now part of the Laurel-Snow State Natural Area, which is just northeast of Dayton. Much of DCIC's land remains blackened by coal dust.

Sources

Broyles, Bettye J. 1991. *History of Rhea County, Tennessee*. Collegedale, TN: College Press, 107.

Chappuis, Erin, *et al.* 2018. Reparations by Dayton Coal and Iron Company. *Rhea Notes*, 23 (1), 1–2.

Johnson, Reed. 2012. How a distance of 2' shaped the history of Dayton. *Rhea Herald News*, May 9, A1-A2.

Miners entombed in a Tennessee mine. 1895. *Pittsburgh Commercial Gazette*, December 19, 1.

Dayton, Tennessee
ca. 1820

Dayton, Tennessee, is a town about 35 miles northeast of Chattanooga in southeastern Tennessee. The area was settled around 1820 as Smith's Crossroads and was renamed "Dayton" (after Dayton, Ohio) in 1877. Dayton's population (220 by 1880) grew rapidly when the Southern Railway came through town. Dayton was incorporated in 1903, at which time its primary industries were agriculture and the production of coal and pig iron by Dayton Coal and Iron Company (DCIC). DCIC's collapse in 1913 cratered the local economy and prompted local businessmen to search for ways to revive the area's economic fortunes.

In 1925, Dayton was only one mile long and five blocks wide when it hosted the biggest event in its history: the Scopes Trial. Reporters and writers in Dayton for the trial expressed vastly different opinions about the town. One writer described Dayton as being "drunk on religious excitement" (Stone, 1943); H.L. Mencken initially praised Dayton, but soon thereafter condemned it and its residents. Conversely, W.O. McGeehan admitted that he "found indigestion," but—after another reporter had ridiculed the town's "morons"—added that "Dayton is a small town with a Main street, but I found there beauty, courtesy, and intelligence.... The percentage of moronism in Dayton ... is no higher than it is in New York City or in any other city.... If I were fishing for morons, I should throw my dragnet at Forty-second and Broadway, New York, rather than at Main and Market streets, Dayton" (McGeehan, 1925).

On May 16, the *Chattanooga News* proclaimed that Dayton had become "the center of the world struggle between Modernists and Fundamentalists." Local businessmen hoped that John Scopes' trial would attract attention and stimulate the local economy, but the hoped-for economic boom never happened. Instead, many people began referring to Dayton derisively as "Monkey Town,"[13] prompting John Scopes' friend Whitney Morgan to admit that "it wasn't 90 days until we realized that [the Scopes Trial] was the worst publicity we could get." Today, however, many of Dayton's residents have embraced their history; "Monkey Town" is prominent on a town mural and is the name of a popular restaurant and brewery.

In 2022, Dayton had approximately 7,400 residents. Its primary industries produce furniture, clothing, and car tires, and its largest university is Bryan College, a memorial to Scopes' prosecutor William Jennings Bryan. People from around the world come to Dayton to learn about the Scopes Trial and photograph the Rhea County Courthouse, where the trial occurred.

Sources

Allem, Warren. 1959. Background on the Scopes Trial at Dayton, Tennessee. Master's Thesis, University of Tennessee, Knoxville, Tennessee, 100.

McGeehan, William O'Connell. 1925. Why pick on Dayton? *Harper's Magazine*, 151 (October), 623–627.

Stone, Irving. 1943. *Clarence Darrow for the Defense*. Garden City, NY: Doubleday, 436.

Epperson, Susan
b. 1941

Susan Epperson was born on May 21, 1941, in Little Rock, Arkansas. Susan's father taught biology at the College of the Ozarks, and she often accompanied him to work to see the plants and animals in his classroom. Susan earned a B.S. in biology in 1962, followed by an M.S. in zoology two years later from the University of Illinois in Urbana. In Fall 1964, Susan began working as a 10th-grade biology teacher at Little Rock's Central High School. There, she learned that Arkansas, like Tennessee, banned the teaching of human evolution in its public schools. The Arkansas law, which had been passed by a public referendum in 1928, was why most biology teachers at Susan's school skipped the topic in their classes.[14]

In 1965, Arkansas biology teacher Susan Epperson challenged her state's Butler Act–like law banning the teaching of human evolution in its public schools. Three years later, the U.S. Supreme Court ruled in *Epperson v. Arkansas* that such laws are unconstitutional. This photograph shows Susan in her attorneys' office in June 1967 while preparing for her case to be heard before the U.S. Supreme Court (Susan and Jon Epperson).

Epperson faced a dilemma: If she obeyed the law and did not teach evolution to her students, she would be committing educational malpractice and cheating her students. However, if she did teach evolution, she would be knowingly violating the law and risk being fired. When the Arkansas Education Association (AEA) asked Susan to test the law, Susan—encouraged to do so by her husband, Jon—agreed "because of my concept of my responsibilities as a biology teacher and an American citizen." Epperson's case was the first since the Scopes Trial to challenge an antievolution law.

When Epperson's lawsuit became public, the *Arkansas Gazette* praised Epperson, as did John Scopes. However, Arkansas Governor Orval Faubus announced that he "was not ready to repeal the Bible," and Arkansas Attorney General Bruce Bennett denounced evolution as atheistic as he assured Arkansas's citizens that he considered Epperson's lawsuit an attack on the Bible that he would defend.

Epperson's first trial, which the press promoted as "Scopes Trial II" and "The New Monkey Trial," was held on April 1, 1966. Chancery Judge Murray Reed, hoping to avoid the shenanigans of the Scopes Trial, announced in court that "the only question here is a constitutional question of law and it is a serious one" (Lewis, 1966). In doing so, Reed banned questions about Susan's personal beliefs and the scientific validity of evolution. Faced with these restrictions, Bennett invoked the majoritarian argument used by William Jennings Bryan in Dayton, asking, "Are the rights of the majority to be ignored in order that Mrs. Epperson's desire to teach evolution be gratified? ... Have the people no right or power, by majority vote, to forbid the teaching of a theory in which they do not believe, to which they violently object, which has no basis in fact, and which is detrimental to the spiritual and mental well-being of their children?" Bennett then ended his argument by invoking Abraham Lincoln's words from Gettysburg, adding, "If not, then this is no longer a government of the people, by the people, and for the people." Epperson's hearing took less than two hours, during which the carnival atmosphere of the Scopes Trial never materialized.

On May 27, 1966, Judge Reed issued a nine-page decision declaring Arkansas's anti-evolution law to be unconstitutional. According to Reed, the Arkansas law violated constitutional protections of free speech and "hinder[s] the quest for knowledge, restrict[s] the freedom to learn, and restrain[s] freedom to teach" (Moore, 1998a). Although Reed's decision again made it legal to teach evolution in Arkansas, Governor Faubus immediately condemned Reed's handling of the case and urged the Arkansas Supreme Court to reverse Reed's decision (Antievolution law is good, 1966). The justices must have been listening, for on June 5, 1967, just before adjourning for their summer recess, the Arkansas Supreme Court issued a two-sentence decision stating that the Arkansas antievolution law was "a valid exercise of the state's power to specify the curriculum in its public schools." It was again a crime to teach human evolution in Arkansas's public schools. A disappointed Epperson lamented, "I'd hoped that our state would wash its own dirty dishes, but it didn't" (Billington, 1968).

Evolution Reaches the U.S. Supreme Court

Epperson's case—now known as *Epperson v. Arkansas*—was then appealed to the U.S. Supreme Court, where oral arguments on October 16, 1968, lasted only 35 minutes. When Justice Thurgood Marshall saw that the Arkansas Supreme Court's opinion comprised only two sentences, he suggested that the U.S. Supreme Court could dispense its opinion in only one.

On November 12, 1968, the U.S. Supreme Court—citing "the celebrated Scopes case"—ruled unanimously that the Arkansas law banning the teaching of human evolution was unconstitutional because (1) the law originated because "a particular religious group considers the evolution theory to conflict with the account of the origin of man set forth in the Book of Genesis" and was therefore not religiously neutral, and (2) "a State's right to prescribe the public school curriculum does not include the right to prohibit teaching a scientific theory or doctrine for reasons that run counter to the principles of the First Amendment." By 1970, all laws banning the teaching of human evolution in public schools had been overturned.[15]

Epperson was pleased that biology teachers could again "teach the full subject of biology with an open and clear conscience." The *Arkansas Gazette*, which commended the courts for rescuing citizens "from a law of ignorance," praised Epperson for removing "an obstacle to quality education that was rooted in ignorance and perpetuated through 40 years of legislative indifference" (An overdue rescue, 1968). *Epperson v. Arkansas* was later applied to a variety of other legal issues, including those involving censorship of textbooks, "release time" from public schools for religious training, Bible readings in public schools, sex education, and the academic freedom of teachers.

In Dayton, Tennessee, however, where the Scopes Trial had occurred more than 40 years earlier, citizens insisted that they would "convict Scopes still" (Dayton says, 1968).

Epperson Meets Scopes

On January 6, 1969, Susan Epperson enjoyed lunch with John Scopes at a Holiday Inn in Shreveport, Louisiana, where Scopes told Epperson "that his not being [at my trial] did not mean he didn't support our efforts" (Moore, 1998b). (Scopes, who then was 68 years old, had declined several offers by newspapers to attend Epperson's trial.) Scopes had received more letters about Epperson's case than had Epperson, and he knew that the struggle for academic freedom would go on. Nevertheless, he was pleased with the Supreme Court's decision, noting, "This is what I've been working for all along.... I'm happy about the decision. I thought all along—ever since 1925—that the law was unconstitutional" (a Darwinian ban, 1968).

After raising her family, Epperson resumed her career as a biology teacher, but was never again involved significantly in the evolution-creationism controversy. In 2022, Susan and her husband Jon were living in Colorado, where Susan continues to promote the compatibility she sees between faith and science.

Sources

Antievolution law is good, Faubus says. 1966. *Arkansas Gazette*, April 5.
Billington, Joy. 1968. Susan Epperson is guinea pig in "Monkey Trial" reenactment. *The Sunday Star* (Washington, D.C.), December 1, H2.

On January 6, 1969, Susan Epperson enjoyed lunch with John Scopes in Shreveport, Louisiana. Scopes had followed Susan's case, but had not gotten involved because he did not want to be a distraction (Jerry Tompkins).

A Darwinism ban in Arkansas ends. 1968. *New York Times*, November 13, 1, 4.
Dayton says it'd convict Scopes still. 1968. *Arkansas Gazette*, November 14.
Lewis, Bill. 1966. Spectators told to behave; trial no April Fool's joke to judge, woman's lawyer. *Arkansas Gazette*, April 2, 2A.
Moore, Randy. 1998a. Creationism in the United States. III. The ban on the teaching of evolution reaches the U.S. Supreme Court. *American Biology Teacher*, 60 (9), 650–661.
______. 1998b. Thanking Susan Epperson. *American Biology Teacher*, 60 (9), 642–646.
An overdue rescue from a law of ignorance. 1968. *Arkansas Gazette*, April 24, 6A.

Eugenics

In 1883, Sir Francis Galton—a statistician and Charles Darwin's half-cousin[16]—coined the term *eugenics*, which he defined as "the improvement of the human race by better breeding" (Allen, 2019). In subsequent decades, many scientists began appealing to eugenics to disguise their social biases as objective, enlightened, facts-of-nature based on evolution. Eugenicists misused evolution and other data to promote biological determinism portraying social inequalities based on gender, race, and class as unalterable facts based on innate and inferior genetic endowments.

In the 1920s, the growth of the fundamentalists' antievolution movement coincided with the growing acceptance of eugenics. Eugenicists demanded restrictions on immigration and marriage, as well as the segregation and forced sterilization of "defectives" to save society. Between 1907 and 1935, some 30 states passed laws that resulted in the forced sterilization of 65,000 people (Allen, 2019). To advocates of eugenics, racial hierarches and social inequalities were the result of biological facts, not bigotry. Although eugenics was promoted as an ethical duty ostensibly about improving the genetics of populations and producing a social utopia, it disproportionately targeted the disabled, mentally ill, and minorities. Eugenics, a form of applied hereditary law, became a byword for Social Darwinism, racism, and discrimination.

In April 1922, the American Association for the Advancement of Science (AAAS) created its "Committee on Evolution" to defend the teaching of evolution in public schools. The committee, which often consulted with John Scopes' attorneys before and during the trial, had three members: Charles Davenport, Henry Osborn, and Edwin Conklin. All three were prominent scientists and eugenicists who contributed to the racism of physical anthropology by bending scientific findings to fit their social biases:

Charles Benedict Davenport (1866–1944), a disciple of Galton who founded the Eugenics Record Office (1910) and worked with Nazis before and during World War II, believed that eugenics was an applied form of heredity. He demanded mass compulsory sterilization and incarceration of the "unfit" because "where the life of the state is threatened, extreme measures may and must be taken" (Davenport, 1913).

Henry Fairfield Osborn (1857–1935), who at the time of the Scopes Trial was "second only to Albert Einstein as the most popular and well-known scientist in America" (Regal, 2002), hosted the Second and Third International Congresses of Eugenics in New York City. Osborn endorsed (and wrote prefaces for) Madison Grant's *The Passing of the Great Race* (1916), which Adolf Hitler (1889–1945) praised as "his Bible" because it advocated eliminating the weak and unfit. To Osborn (1922), "only the best has the right to survive."

Edwin Grant Conklin (1863–1952) conflated race, eugenics, and evolution while claiming, "Natural selection must be replaced by intelligent human selection" (Maynard, 2009). According to Conklin, evolution operated according to a eugenic natural law that sorted races hierarchically, and humans needed to intervene to preserve this order. Specifically, Conklin wanted to forcibly sterilize defective people to preserve the purity and superiority of the White race; he also complained that the unfit "are preserved by charity," meaning that "natural selection, the great law of evolution and progress, is set to naught" (Conklin, 1915). Darrow begged Conklin to come to Dayton (e.g., "I feel we must have you") and the ACLU promised that Conklin and Henry Osborn would be there to support Scopes. Conklin and Osborn cited health concerns for not coming to Dayton, but both travelled across the country later in the summer for other reasons. Although Conklin and Osborn had often ridiculed Bryan's views about evolution, they ducked their chance to confront Bryan at Dayton.

Osborn, the "Chair of the Scientific Committee to Defend John Thomas Scopes," may have refused to meet Bryan face-to-face in Dayton because he was uncomfortable with Darrow's opposition to American policies about eugenics (Darrow, 1926; Davis, 2005).

Eugenics was prominent in George Hunter's *A Civic Biology: Presented in Problems*, the state-approved textbook that John Scopes used when he taught biology in Dayton. Scopes apparently covered the material, for he asked his students about the textbook's eugenics-related material on an exam that he gave on April 28. Although Hunter's book was discussed extensively at the Scopes Trial, there was no mention of its advocacy of eugenics. The replacement for Hunter's book, Peabody and Hunt's *Biology and Human Welfare* (1924), did not discuss evolution, but to promoted eugenics as "a great movement" and claimed that "any permanent improvement of the human race can only come as a result of better heritage."

Darrow had earlier urged authorities to "chloroform unfit children," but he opposed American eugenic policies. In his first public speech after being officially chosen to defend Scopes, Darrow linked eugenics with Bryan, noting, "When they talk about changing man by eugenics, I want to know who is going to do the changing. I'll bet it will be the Bryans" (Darrow sneers, 1925). However, some of the scientists brought to Dayton by Darrow and the defense team staunchly supported eugenics. For example, biologist Maynard Metcalf—the only scientist allowed to testify in Dayton from the witness stand—believed it was "criminal" and "heinous" to produce children who "will be an injury to society," and denounced charities and social workers who helped those people reproduce as "accessory to the crime" (Metcalf, 1916). Similarly, zoologist Horatio Hackett Newman promoted eugenics so "that man might control his own evolution and save himself from racial degeneration." His popular textbook, *Evolution, Genetics, and Eugenics*, advocated limiting immigration, restricting marriage, segregating defectives, and forcibly sterilizing defective people (Newman, 1932). William Goldsmith (1888–1955), a biologist who came to Dayton but did not testify, urged society to curb reproduction of the "unfit" or "there is grave danger of the future society being engulfed by degeneracy … the problem … [is] eliminating the unfit by refusing them birth" (Goldsmith, 1922).

David Starr Jordan (1851–1931), a prominent biologist who raised money for Scopes' defense and graduate work, believed that racial decay had resulted from "the survival of the unfit" (Jordan, 1902). Jordan dismissed his racism and support of eugenics as being based on common sense, noting, "To say that one race is superior to another is merely to confirm the common observation of every intelligent citizen." After the Scopes Trial, Jordan—who strongly supported the forced sterilization of disabled people and others—became a founding member of the Human Betterment Foundation, which was dedicated to "sterilization for human betterment" and which inspired the eugenics of Nazi Germany. Thanks to Jordan and other prominent scientists, the United States led the world in forced sterilizations. Germany did not implement such laws until the Nazis came into power in 1933 (Burns, 1953; Allen, 2019).[17]

By 1923, eugenicists were claiming that eugenics policies were improving society, including by reducing crime. Five years later, more than 350 U.S. universities were teaching eugenics, and 70 percent of high-school biology textbooks endorsed eugenics. U.S. eugenics laws, which were infamously endorsed by the U.S. Supreme Court in *Buck v. Bell* (1927), inspired Nazis to justify their own programs of sterilization and, later, extermination. However, in 1935, a panel convened by the Carnegie Institute concluded that most of the work by Davenport's Eugenics Record Office lacked scientific merit and, within a few years, some college biology textbooks began to reject eugenics. Although eugenics programs in the United States faded after they were discredited by Nazi eugenic programs, the topic remained in America's most popular biology textbook—Truman Moon's *Modern Biology*—through its 1963 edition, and in Claude Villee's popular college textbook *Biology* through its 1967 edition.

By the 1930s, most prominent geneticists had rejected eugenics.[18] In the United States, the eugenics movement was replaced by more subtle influences and pressures such as pre-birth genetic counseling and arguments involving the medical and social costs of caring for disabled people. The 2010 movie *Alleged*, a "romantic drama set against the landmark Scopes 'Monkey Trial' of 1925," linked evolution with eugenics through Hunter's *A Civic Biology* and with Abigale, a "half breed" who is targeted for sterilization after scoring poorly on an IQ test. Darrow helped stop Abigale's sterilization.

Sources

Allen, Garland E. 2019. Commentary: Heredity and environment in early 20th century genetics: Major Leonard Darwin's advice to eugenicists. *International Journal of Epidemiology*, 48 (2), 358–361.

Burns, Edward McNall. 1953. *David Starr Jordan: Prophet of Freedom*. Redwood City, CA: Stanford University Press.

Conklin, Edwin Grant. 1915. Value of negative eugenics. *Journal of Heredity*, 6 (12), 538–540.

Darrow sneers at efforts to change man by eugenics. 1925. *Atlanta Constitution*, June 11, 9.

Davenport, C.B. 1913. Fit and unfit matings. *Journal of Proceedings and Addresses of the Fifty-first Annual Meeting, National Education Association of the United States*, July 5–11, 773–783.

Davis, E.B. 2005. Science and religious fundamentalism in the 1920s. *American Scientist*, 93, 253–260.

Goldsmith, William M. 1922. *The Laws of Life: Principles of Evolution, Heredity and Eugenics*. Boston: Gorham Press, 407–408.

Jordan, David Starr. 1902. *The Blood of the Nation: A Study of the Decay of Races Through the Survival of the Unfit*. Boston: American Unitarian Association.

Maynard, W. Barksdale. 2009. The Darwin of Guyot Hall. *Princeton Alumni Weekly*, February 11.

Metcalf, Maynard M. 1916. Evolution and man. *Journal of Heredity*, 7 (8), 356–364.

Newman, Horatio Hackett. 1932. *Evolution, Genetics, and Eugenics*, Third Ed. Chicago: University of Chicago Press, 441–444.

Osborn, Henry Fairfield. 1922. Evolution and religion. *New York Times*, March 4, 91.

Peabody, James Edward, and Arthur Ellsworth Hunt. 1924. *Biology and Human Welfare*. New York: Macmillan, 6, 463–464, 547–548.

Regal, Brian. 2002. *Henry Fairfield Osborn: Race and the Search for the Origins of Man*. Farnham, UK: Ashgate, xii.

Expert Witnesses

In the weeks before the Scopes Trial, the prosecution and defense searched for experts to bolster their arguments and counter their opponents' claims. When Bryan could find no prominent scientists to support his claims, he asked fundamentalist leaders such as Frank Norris and John Roach Straton to come to Dayton. They promised to be there, but none showed up. The defense had similar problems; the ACLU assured reporters that prominent scientists such as Edwin Grant Conklin and Henry Fairfield Osborn would be in Dayton, but none showed up. The scientists who did come to Dayton for the trial admitted to being "of lesser standing" and "second-string scientists as to age and reputation" (Curtis, 1956, 1965).

On the afternoon of Wednesday, July 15, zoologist and Congregationalist layman Maynard Metcalf testified from the witness stand that Earth is ancient and evolution "is a fact ... that is perfectly and absolutely clear." The next morning, Stewart's objection to Metcalf's continuing testimony triggered a day-long debate about the admissibility of the defense team's expert witnesses. On Friday morning, Judge Raulston ruled that scientists' testimonies were inadmissible, but allowed their statements to be read into the court record for possible use during an appeal. Some of the defense team's experts and their affidavits were as follows:

Maynard Metcalf (1868–1940), a zoologist doing private research at Johns Hopkins University at the time of the trial: "Teaching in any field that deals with living things is disgracefully, yes, criminally, inadequate if it omits emphasis upon evolution.... The fact of evolution—of man, of all living things ... is a fully established as the fact that the Earth revolves around the Sun." Metcalf was the only scientist to testify from the witness stand in the trial.

Horatio Hackett Newman (1875–1957), a zoology professor from the University of Chicago: Fossil evidence supports the evolution of humans.

Jacob Lipman (1878–1939), a professor of agricultural chemistry and director of the New Jersey Agricultural Experiment Station: The Butler Act deprives students of "the facts of human development." Without evolution, "agriculture could not provide an effective service to mankind."

Fay-Cooper Cole (1881–1961), an anthropology professor from the University of Chicago: Fossils and "useless" organs are evidence of evolution. Anthropoid apes resembled humans.

Winterton Curtis (1875–1966), a zoologist from the University of Missouri: "The historical fact of evolution has been attested by overwhelming evidence."

Wilbur Nelson (1890–1969) was State Geologist of Tennessee: It would be impossible to study or teach geology without invoking evolution.

Herman Rosenwasser (1875–1956), a rabbi and Hebrew linguist from San Francisco who came to Dayton uninvited and unannounced: "Genesis itself teaches that man is descended from a lower order of animals.... If the Hebrew Bible were properly translated and understood, one would not find any conflict with the theory of evolution."

Kirtley Mather (1888–1978), a geologist and Baptist deacon from Harvard University: Geologists accept evolution; science and the Bible do not necessarily conflict. Mather testified "not because of my love for science, but because of my love for science ... [the Butler Act] is a serious blow to the cause of religion" (Geology expert, 1925).

Herbert Murkett (1870–1958), a Methodist preacher from Chattanooga: "The Bible properly interpreted does not conflict with the theory of evolution ... [the Bible's creation-story] does not tell how man was made, but that he was made from dust."

Charles Judd (1873–1946), director of the Department of Education at the University of Chicago: It is "impossible to obey the law without seriously depriving teachers in training of a proper view" of human mental development.

Shailer Mathews (1863–1941), dean of the University of Chicago's School of Divinity: The Genesis creation story "is no more denied by evolution than it is by the laws of light, electricity and gravitation." Before the trial began, Mathews defined modernists as Christians "who accept the results of scientific research as data with which to think religiously" and denounced the Butler Act as "one of the most dangerous pieces of legislation every passed" (Science and fundamentalism, 1925).

Scientists invited to Dayton by the defense team stayed at the Mansion just outside of town. Shown here on the Mansion's front steps are (back row, left to right) Horatio Newman, Maynard Metcalf, Fay-Cooper Cole, and Jacob Lipman, and (front row, left to right) Winterton Curtis, Wilbur Nelson, and William Goldsmith. Metcalf was the only scientist to testify from the witness stand during the trial (Smithsonian Institution Archives 2008–1148).

Walter Whitaker (1867–1938), an Episcopal priest from Knoxville: After preaching about Jesus for 30 years, "I am unable to see any contradiction between evolution and Christianity."

Although popular horticulturalist Luther Burbank, reporter Watson Davis, defense attorney Arthur Garfield Hays, and a host of newspapers promised that these witnesses would "unroll the evidence" and educate the public about evolution, most coverage of the Scopes Trial focused on personalities and sideshows. Between May and July, only the newspapers' summaries of Maynard Metcalf's testimony stressed on evolution. Otherwise, nearly four-times more stories emphasized unchallenged opinions than facts about evolution (Parks, 2015).

After the trial, the experts went home, where they were often greeted by hate mail, attempts to have them fired, and larger enrollments in classes that they taught (Curtis, 1965). When the Tennessee Supreme Court set aside Scopes' conviction on appeal, the justices did not mention the expert witnesses' statements.

Sources

Curtis, Winterton C. 1956. *Fundamentalism versus Evolution at Dayton, Tennessee: Abstracts from the Autobiographical Notes of Winterton C. Curtis*. Columbia, MO: Winterton C. Curtis, 57.

______. 1965. The evolution controversy. Pp. 74–85 in Tompkins, Jerry R. *D-Days at Dayton: Reflections on the Scopes Trial*. Baton Rouge: Louisiana State University Press.
Geology expert enroute to trial. 1925. *Knoxville Journal*, July 19, 6A.
Parks, Perry. 2015. Summer for the scientists? The Scopes Trial and the pedagogy of journalism. *Journalism and Mass Communication Quarterly*, 92 (2), 444–467.
Science and fundamentalism. 1925. *Chicago Tribune*, April 1, 8.

First United Methodist Church
Est. 1833

Dayton's First United Methodist Church was established on August 31, 1833, when the Methodist Episcopal Church bought five acres of land near Smith's Crossroads. After the Civil War, the church was attended by Democrats and Confederate sympathizers. Republicans and Union loyalists attended First Avenue Methodist Episcopal Church (est. 1888).

The pastor of First United Methodist Church during the Scopes Trial was the Rev. Charles R. Jones. On July 12, 1925, William Jennings Bryan—who had not yet spoken in the courtroom during the trial—preached to an overflow crowd at the church's morning service. Sitting on the front row were Judge John Raulston and his family. Two weeks later, on July 26, Bryan drove to Dayton from Chattanooga and made an unannounced appearance at the church, where he led a prayer. This was Bryan's last public appearance. After Bryan died that afternoon, Jones officiated a memorial service for Bryan at the home of F.R. Rogers just down Market Street.

On April 23, 1968, Methodist Episcopal Church changed its name to First United Methodist Church. On July 24, 1994, as part of the festivities associated with the Scopes Trial Play and Festival, the church recreated the July 26, 1925, service attended by Bryan. Today, the historical significance of First United Methodist Church (South) is designated by a Scopes Trial Trail plaque near the main entrance.

Source

History of First United Methodist Church. Obtained from member Donna Taylor Reed in July 2017.

The Fundamentals: A Testimony to the Truth
1910–1915

The Fundamentals: A Testimony to the Truth was a series of 12 paperback books published between 1910 and 1915 meant "to strengthen earnest believer … warn and re-establish in the truth those who are wavering in their faith, and lead repentant sinners to conviction of sin and the faith in the Lord Jesus Christ." The books were paid for by Lyman Stewart (1840–1923) and Milton Stewart (1838–1923), the founders of Union Oil and the Bible Institute of Los Angeles (now Biola University).[19] By the time the last volume appeared in 1915, almost three million books were sent to preachers, Christian youth organizations, missionaries, and others.

Before 1920, fundamentalists may not have been comfortable with evolution, but the topic was not a central tenant of their crusade. Thus, of the 90 articles in

The Fundamentals the generally supported Biblical inerrancy, only three addressed evolution. For example, Scottish theologian James Orr (1844–1913) promoted a God-driven theistic evolution while claiming that "evolution, in short, is coming to be recognized as but a new name for 'creation' … here, again, the Bible and science are felt in harmony" (*The Fundamentals*, 1910). Several authors stressed the importance of the supernatural; for example, geologist and Calvinist minister George Frederick Wright, who believed that God is the source of variation that drives natural selection, claimed that the "evidence for evolution … does not begin to be as strong as that for the revelation of God in the Bible," to which Howard Crosby added, "The Word is supernatural … woe to the preacher who leaves the supernatural for the natural" (*The Fundamentals*, 1912). None of the articles in *The Fundamentals* focused on the teaching of evolution.

The Fundamentals were not cited in the Scopes Trial transcript, were largely ignored by academic and denominational presses, and did not stop the advance of modernism or evolution in American society. Nevertheless, the books became a symbolic landmark for the emerging fundamentalist movement that produced the Scopes Trial.

Sources

The Fundamentals: A Testimony to the Truth. 1910–1915. Chicago: Testimony Publishing Co.

Godsey, John Luke
1875–1932

John Luke Godsey was born on February 9, 1875, and by early 1925 was a prominent attorney in Dayton. Following the meeting at Robinson's Drug Store on May 5, 1925, that ultimately produced the Scopes Trial, Godsey agreed to help defend Scopes. On May 9, Godsey and John Randolph Neal represented Scopes at a preliminary hearing, and on June 13 Godsey argued (unsuccessfully) that the charges against Scopes should be dropped because the Butler Act was unconstitutional. Although Godsey believed that evolution is compatible with the creation story in the Bible, he also believed that the Butler Act was a "foolish law" that "would not only put the church into the schools, but it would make the Bible a part of the state statutes … [by] establishing a state church." He went on, asking "what does the Biblical story mean? Christian ministers themselves differ violently over its meaning. Whose construction will the court take.... By common consent of qualified persons, [the Butler Act] would seriously cripple the teaching of nearly all the sciences in the public schools. What we want our children to learn in the schools is the truth, regardless of any particular religious doctrine that may be affected.... As a parent, I have the right to demand that my child be taught the truth, and when that is denied, my privileges are denied" (Anderson, 1925).

On the first day of the trial, Godsey—who was described as "a substantial, old-fashioned local lawyer"—resigned from the defense team because he was "too busy." However, newspapers reported that he had resigned "owing to a nervous breakdown incident to the work of the trial" (Evolution is a home study, 1925), and Darrow claimed that Godsey "ran away from the case in fear and trepidation. The sentiment of the town and the State was more than he could face" (Darrow, 1932). Other accounts claimed that Godsey was considering running against Judge Raulston in the next election and did not want to be associated with the trial, evolution, or Clarence Darrow, adding, "No wonder Godsey decided the water was getting cool" (Kinsley, 1925). Whatever the reason, Godsey admitted, "I wanted to get out of the case and I did. I have withdrawn. My health has not been as good as it might be, and I am out" (John L. Godsey, Scopes first lawyer, 1925). Godsey, who did not attend the trial, was replaced by F.B. McElwee, a former student of John Randolph Neal. McElwee played a minor role in the trial.

Godsey, an incorporator of the Bryan Memorial University Association, died in Dayton on June 27, 1932, and is buried in Dayton's Buttram Cemetery.

Sources

Anderson, Paul Y. 1925. The drugstore argument that swelled into a national debate. *St. Louis Post-Dispatch*, May 31, 17, 20.
Darrow, Clarence. 1932. *The Story of My Life*. New York: Charles Scribner's Sons, 249.
Evolution is a home study. 1925. *Lincoln (NE) Journal Star*, July 10, 1.
John L. Godsey, Scopes' first lawyer, withdraws from evolution case. 1925. *Chattanooga Daily Times*, July 11, 1.
Kinsley, Philip. 1925. Court may bar Scopes experts. *Ft. Worth Star-Telegram*, July 12, 1, 6.

Grand Jury Testimony

When John Scopes was arrested for violating the Butler Act in early May 1925, Judge John Raulston was not scheduled to hold court in Rhea County until Fall. However, to expedite the case and keep other towns from upstaging Dayton with a test-case of their own, Judge Raulston convened the grand jury on May 25. At that meeting, Raulston posed for pictures, read the statute and the first chapter of Genesis, and announced that he would consider any violation of the Butler Act to be "a high misdemeanor" because it involved "the evil example of the teacher disregarding constituted authority." Raulston added, "It has always been and is now one of the great passions of my life to ascertain the truth about all matters, especially relative to God, but I am not so much exercised over the question as to whence man cometh as I am to whither he goes" (Judge's own views, 1925).

Seven of Scopes' students were called by the grand jury, but not all of them testified. James Benson, a native of Dayton, Tennessee, was a senior in John Scopes' physics class at Rhea Central High School in Spring 1925. Benson testified that in early April, when a classmate asked Scopes about Darwin's theory of evolution, "Scopes said that Darwin's theory of evolution was that 'life started on earth with the one-celled plants and animals and gradually developed up to the higher and more complex forms of animal and plant life, and man was the climax of this development'" (Of Monkeys and Men, 2021). Benson added that Scopes' comments were irrelevant to the class (Moutoux, 1925). Although William Jennings Bryan made much during the trial of the textbook that Scopes used in his biology class, Benson testified that Scopes discussed evolution independent of that textbook in his physics class.

Jack Hudson, another student who was subpoenaed (and who had played on Scopes' football team), told a reporter, "It's a funny mess. We'd be the last ones in the world to do a thing against Prof. Scopes and yet they picked on us to testify against him before the grand jury and get him indicted. But he wanted us to do it—he wanted to get pinched." Hudson and other students later warned, "If Scopes doesn't come back to this school, they'll be trouble from all the kids" (Sumner, 1925). Another of Scopes' students told a reporter that "someone in [Scopes' biology] class once asked Scopes if he thought man originated from a monkey," to which "Scopes half grinned and said, 'Probably man did'" (In Scopes Trial figures recall old memories, 1960).

On May 25, after deliberating for an hour, the grand jury indicted Scopes (Indictment 5231). However, the term of that grand jury had expired before the trial began, so Judge Raulston convened another grand jury during a "Special Term of the Court" on the first day of Scopes' trial. After hearing testimony from some of Scopes' students (Howard Morgan, James Benson, and Jack Hudson), the newly empaneled grand jury re-indicted Scopes (No. 5232) for violating the Butler Act on April 24, a day that Scopes had been sick and missed class.

Scopes' students Howard Morgan and Bud Shelton testified during the trial. However, Benson and some of Scopes' other students (e.g., Orville Gannoway, Morris Stout) who had been subpoenaed to appear were dismissed before testifying because their testimony was going to be "substantially the same as the Shelton boy."

After the trial, Benson earned a B.S. degree from East Tennessee State College in chemistry, and Master and Ph.D. degrees from the University of Tennessee in physics and math. He then worked at Langley Aeronautics Laboratory at what is now NASA, and then for decades worked with Hastings-Raydist, where he registered several patents about vacuum gauges, navigational equipment, and other devices (Dr. James Benson, 1996). Benson died on May 26, 1996, and is buried in Greenlawn Cemetery in Newport News, Virginia. Shelton and Morgan are described elsewhere in this book.

Sources

In Scopes Trial figures recall old memories. 1960. *Knoxville News-Sentinel*, July 17, B-10.
James M. Benson, Scopes Trial witness. 1996. *Daily Press* (Newport News, VA), May 28, 14.
Judge's own views. 1925. *Nashville Banner*, May 25, 5.
Moutoux, John T. 1925. Evolution teacher is indicted. *Knoxville News-Sentinel*, May 25, 1, 8.
Of Monkeys and Men: Public and Private Views of the Scopes Trial. Special Collections, University of Tennessee, Knoxville, Tennessee.
Sumner, Allene. 1925. Scopes begged boys to tell. *Paducah News-Democrat*, July 5, 1.

Haggard, Andrew Pleasant
1862–1932

Pleasant Andrew Haggard was born at Spring Place in Murray County, Georgia, on August 2, 1862. After graduating from the National Normal University in Lebanon, Ohio, he began teaching in Georgia, where he married one of his students, Arrena "Rena" Clark. One of their sons, Wallace, helped prosecute John Scopes. According to Rena, A.P. Haggard was not always pleasant, so he legally changed his name from Pleasant Andrew Haggard to Andrew Pleasant "A.P." Haggard.

The Haggards moved to Dayton in 1885, after which A.P. started several businesses and served as a correspondent for the *Chattanooga Times* newspaper. In 1916, Haggard—who became Dayton's commissioner and most prominent businessman—was one of five incorporators who tried to revive Dayton Coal & Iron Company, but the venture failed

(Broyles, 1991). At the time of the Scopes Trial, Haggard—who was also known as "Mr. Dayton"—was the richest and most prominent person in Dayton; he owned several local businesses, including American National Bank. Newspapers described Haggard as "the outstanding Rhea County leader in practically every public-spirited undertaking" who had "given attention to many interests of [Dayton's] social and education welfare" (Dayton's mayor weds Mrs. Collins, 1925).

Before and during the trial, Haggard received hundreds of letters. He answered only one, from E.M. Clackenburg of Chicago, which asked, "Will you kindly let me know if the legislative body of Dayton has passed an emergency ordinance prohibiting squirrels within the corporate limits of Dayton, during the Scopes trial, as I think the nuts should be allowed to ripen before they are picked, and oblige." Haggard replied: "We regret that on account of the train of cranks and anarchists who followed some of the advocates of evolution to our city, it required all our energies, thought and force to protect our people against their assaults, and for this reason we were unable to consider the ordinance you refer to. No doubt you have already protected the nuts of your city against the ravages of squirrels; now let me suggest that you turn your attention to the anarchists and gunmen who feel more at home and more nearly immune against restriction than they have felt during their sojourn with us" (Mayor Haggard bests Chicago would-be wit, 1925).

Haggard, the "official host of the almost 200 newspaper men and women" who came to Dayton for the trial, was responsible for readying Dayton for the Scopes Trial (Mayor Haggard weds Mrs. Margaret Collins, 1925). Under his direction, workers painted and refurbished the courtroom, brought in more spittoons, built several public bathrooms, installed a barbeque pit and hitching posts for horses on the courthouse lawn, and cleaned the jail and streets in preparation for the hordes of visitors (and bootleggers) expected for the trial. Haggard and Rhea County Sheriff Robert "Bluch" Harris asked Tennessee Governor Austin Peay to send National Guard troops to Dayton to help with security, crowd control, and the "influx of pickpockets, gamblers, and thieves" that would accompany the trial, but Peay denied their request (Dayton asks for state guardsmen, 1925). Undaunted, Haggard and Harris recruited officers from Chattanooga to police the trial. Raulston appointed one of those officers, Kelso Rice, the court bailiff.

When the defense tried to move the trial to a federal court, Haggard fretted that "the city would lose between $400 and $500," which would cover "the price of several septic tanks which the state board of health advised us to install" (Moutoux, 1925). As the trial approached, Haggard repeatedly told reporters, "The charge that Dayton conceived the idea of testing the anti-evolution law as a publicity scheme is absolutely false" and promised that "Dayton will be prepared for the crowds that will attend the trial" (College to have hotel suite, 1925).

When Judge Raulston mentioned on July 17 that "we don't know just how strong [the floor of the courthouse] is," Haggard reassured people that "it certainly is safe … there is nothing at all the matter with it" (Dayton courthouse is safe, 1925). Nevertheless, Raulston, citing the courtroom's "heavily burdened" floor, moved proceedings outside to protect spectators' safety. Scopes later admitted that although Raulston's claim "made a good story; the fact was otherwise: It was the extreme heat that shifted the Court outdoors" (Scopes and Presley, 1967).

When Bryan died in Dayton on July 26, Haggard announced, "The foremost citizen of the world has fallen while the honored guest of our city. Dayton has never been as honored as by his 20 days sojourn in our midst. He fell while in the line of duty in the cause dearest to his heart in all his honorable and noble career. All Dayton mourns his untimely death." Haggard declared the next day to be a holiday on which Daytonians would "close all places of business, display their flags at half-mast, and refrain from all avocations as a token of our sorrow for his untimely death" (Dayton issues proclamation, 1925).

Three months later, Haggard married Margaret Collins, the sister of Sue and Herbert Hicks. Haggard's pledge of $40,000 for Bryan Memorial University to build Rena Clark Haggard Hospital was the largest pledge to come from Rhea County.

In 1927, nearby Morgantown bought a $500 tract of land from Cumberland Coal & Iron Company and opened the Haggard Grammar School, which was named in honor of Haggard (Guffey, 2014). However, Haggard's American National Bank failed in 1928 and Haggard went broke. His health deteriorated and by 1932 he had moved into the home of F.E. Robinson, his son-in-law. There, in an upstairs bedroom on November 14, 1932, the 71-year-old Haggard—facing an upcoming federal indictment for misusing his bank's money—slashed his throat with a razor. He died two hours later (Name four men by indictments in Dayton case, 1932; Knife slash fatal to A.P. Haggard, 1932).

After a well-attended funeral at the Robinsons' home, Haggard was buried beneath a prominent hilltop tombstone in Dayton's Buttram Cemetery. By 1940, the Haggard Grammar School in nearby Morgantown had dropped Haggard's name (Guffey, 2014).

Sources

Broyles, Betty J. 1991. *History of Rhea County, Tennessee.* Dayton, TN: Rhea County Historical and Genealogical Society, 315.
College to have hotel suite at evolution trial. 1925. *Nashville Tennessean*, May 27, 1.
Dayton asks for state guardsmen. 1925. *Nashville Banner*, June 20, 1.
Dayton courthouse is safe for crowd. 1925. *Chattanooga Daily Times*, July 17, 1.
Dayton issues proclamation. 1925. *Nashville Banner*, July 27, 1.
Dayton's mayor weds Mrs. Collins. 1925. *Chattanooga Daily Times*, October 9, 7.
Guffey, Pat. 2014. Haggard Grammar School. Accessed November 1, 2020, at https://www.rheaheraldnews.com/lifestyles/article_180e1a3e-b6ae-11e3-88bf-0019bb2963f4.html.
Knife slash fatal to A.P. Haggard. 1932. *Chattanooga Daily Times*, November 14, 1.
Mayor Haggard bests Chicago would-be wit. 1925. *Chattanooga News*, July 20, 8.
Mayor Haggard weds Mrs. Margaret Collins. 1925. *Chattanooga News*, October 8, 5.
Moutoux, John. 1925. Would have evolution tried in federal court without witnesses. 1925. *Johnson City (TN) Staff-News*, July 3, 1.
Name four men by indictments in Dayton case. 1932. *Chattanooga Daily Times*, November 16, 10.
Scopes, John T., and James Presley. 1967. *Center of the Storm: Memoirs of John T. Scopes.* New York: Holt, Rinehart and Winston, 167.

Haggard, Wallace Callahan
1900–1971

Wallace Callahan Haggard was born on January 22, 1900, to Andrew Pleasant Haggard and Arrena Clark Haggard. After graduating from McCallie School in Chattanooga, Wallace earned a bachelor's degree in civil engineering at Vanderbilt University, where he played on the football team. He then attended Vanderbilt Law School

On the afternoon of July 7, William Jennings Bryan arrived in Dayton, where he was met by Wallace Haggard (far left) and the prosecution team (minus Tom Stewart) and escorted to the home of pharmacist F.R. Rogers. Bryan and his entourage stayed at the Rogers's home throughout and after the trial. This photograph, taken just after Bryan arrived, shows the defense team on the front steps of the Rogers's home, where Bryan would die 19 days later. Left to right: Wallace Haggard, Gordon McKenzie, Rogers, Bryan, Sue Hicks, Herbert Hicks, and Ben McKenzie (RCHGS).

and returned to Dayton to practice law. In the Fall of 1922, Wallace coached the Rhea Central High School "Yellow Jackets" football team in a winless, scoreless season.

Wallace, attorneys and brothers Sue and Herbert Hicks, and Rhea County School Superintendent, Walter White were the first people associated with John Scopes' prosecution. To plan their strategy, Haggard and the Hicks brothers—whom trial instigator George Rappleyea claimed were the worst fundamentalists "in the world"—met with William Jennings Bryan in early June in Nashville. A few weeks later, they met with him again in Atlanta. That meeting was the last meeting of the prosecution before the trial; all of their other planning was done by mail. As secretary of Dayton's Progressive Club, Wallace also helped arrange the banquets that welcomed Bryan and Darrow to town. Wallace, like most of the other prosecutors, served without pay in the trial.

Wallace prepared the prosecution's witnesses for the trial but did not play a significant role in courtroom proceedings. After Clarence Darrow was cited for contempt during the trial, Wallace and others urged Judge Raulston to be lenient, and Chattanooga attorney Frank Spurlock posted the $5,000 bond. When Bryan died in Dayton five days after the trial, Wallace, Gordon McKenzie, and the Hicks brothers watched Bryan's body throughout the night (Associated Press, 1925). Haggard then accompanied Bryan's body to Washington for the funeral.

Wallace later told reporters that fundamentalists "thought they should be allowed to teach the Bible if the theory of evolution was being taught.... They felt they were entitled to some protection of their own beliefs. John Godsey, Sue Hicks, and myself used to sit around and argue about the constitutionality of the [Butler Act] and one day at Robinson's Drug Store ... we talked to John Thomas [Scopes] about it. He volunteered to let them use him to test the law.... We never expected [the case] to gain any attention" (Hardaway, 1960).

After the Scopes Trial, Wallace worked at his father's bank (American National Bank) and Dayton Veneer and Lumber Mills. On November 18, 1932, Wallace Haggard, A.P. Haggard, F.E. Robinson (Wallace's father-in-law), and A.B. Andrews (a former official of the Dayton Hosiery Mills) were indicted for violating banking laws at Dayton's then-closed American National Bank. After the elder Haggard killed himself at Robinson's home, all of the charges were dropped (Tennessee bank case, 1932). Wallace then moved to Americus, Georgia, where he owned Dayton Baskets, which made products of plywood and veneer. Wallace, an incorporator of the Bryan Memorial University Association and William Jennings Bryan University, served for many years as a trustee of the school.

Wallace Haggard died on August 28, 1971, and is buried in Oak Grove Cemetery in Americus, Georgia.

Sources

Associated Press. 1925. Scopes case comrades keep watch over body. *Knoxville News-Sentinel*, July 26, 1.

Hardaway, Sybil. 1960. Dayton key important trophy to Mrs. Scopes. *Chattanooga News-Free Press*, July 22, 4.

Tennessee bank case brings 4 indictments. 1932. *New York Times*, November 19, 8.

Haldeman-Julius, Emanuel 1889–1951

Emanuel Julius was an American socialist and writer born into a family of bookbinders in Philadelphia, Pennsylvania. After working in textile mills, Julius began writing for and editing several newspapers, including his *Appeal to Reason* (1915–1922). When Julius married heiress and aspiring actress Marcet Haldeman, he adopted her surname in hyphenate. The couple started Haldeman-Julius Publications in Girard, Kansas, which became famous for publishing popular "Little Blue Books" featuring radical texts and reprints of classical literature. These 3.5" by 5" books, which were not always blue, initially sold for 5¢ and were the first mass-marketed paperbacks in the United States. The Haldeman-Juliuses sold millions of the books, including several featuring Clarence Darrow (Lee, 2017).

Haldeman-Julius, a quirky, socialist friend of John Scopes' father who was as aggressive in his agnosticism as the fundamentalists were in their faith, drove with his wife to Dayton for the Scopes Trial. Marcet described John Butler as being "full of innate courtesy and kindness." However, her husband reported in the *Haldeman-Julius Weekly* that Scopes was being "lynched" and that Judge Raulston was an "avowed enemy of science and education" and "a tool of the Bryans" who was trying "to crush scientific knowledge and progress" (Superstition versus science, 1925). One of the best-selling "Little Blue Books" (No. 1424) was "The Famous Examination of Bryan at the Scopes Evolution Trial."

After the trial, Haldeman-Julius returned to Kansas, and the following year he bought the last 1,000 copies of the National Book Company's *World's Most Famous Court Trial: Tennessee Evolution Case*, which was the first complete transcript of the Scopes Trial (Hilleary and Metzger, 1925). In 1932, he ran unsuccessfully for the U.S. Senate as a Socialist. In June 1951 he was convicted of income tax evasion, fined $12,500, and sentenced to 18 months in federal prison, but he remained free (on $2,000 bond) while he appealed his case. A month later, on July 31, 1951—one day after his sixty-second birthday—Haldeman-Julius drowned in his swimming pool. He is buried in Cedarville Cemetery in Cedarville, Illinois.

Sources

Hilleary, William, and Oren Metzger, Eds. 1991 [1978, 1925]. *World's Most Famous Court Trial: Tennessee Evolution Case*, Reprint with additions. Dayton, TN: Bryan College.

Lee, R. Alton. 2017. *Publisher for the Masses: Emanual Haldeman-Julius*. Lincoln: University of Nebraska Press.

Superstition versus science in the hills of Tennessee. 1925. *Haldeman-Julius Weekly*, July 18.

Hays, Arthur Garfield
1881–1954

Arthur Garfield Hays was born on December 12, 1881, in Rochester, New York, to wealthy parents in the clothing manufacturing business. Hays was named for the three U.S. presidents during his birth year (i.e., 1881): Charles A. Arthur, James A. Garfield, and Rutherford B. Hayes (without the surname's "e"). Hays graduated from Columbia College in 1902, and in 1905 earned a LL.B. from Columbia Law School. Soon thereafter, he was admitted to the New York bar and, with two former classmates, formed a law firm that represented Germans who were discriminated against because of Germany's opposition to the Allies in World War I. From 1914 to 1915, Hays practiced international law in London, and in 1920 he cofounded the American Civil Liberties Union (ACLU), for which he served as general counsel until his death. Hays's involvement with the Scopes Trial came through his involvement with the ACLU, which placed the advertisement in Tennessee newspapers that ultimately led to the trial.

At the Scopes Trial, Hays—an expert on technical aspects of law who believed that "something is wrong with a faith which is endangered by knowledge"—was the defense team's chief strategist, workhorse, and manager. Hays believed that the trial was "a battle between two types of mind—the rigid, orthodox, accepting, unyielding, narrow conventional mind and the broad, liberal, critical, skeptical, and tolerant mind." On the eve of the trial, Hays—who was in charge of keeping the record in shape for the appeal of Scopes' all-but-certain conviction in Dayton—promised that the trial was "going to be a good education for the people and for the newspapers" (Hays, 1937; Malone glad, 1925).

Hays, the only defense lawyer who reviewed the books brought to Dayton by the Rev. Charles Potter (the defense team's librarian), acknowledged Bryan's heroic status during the trial's early days, but soon thereafter noted that "Darrow's kindliness, charm and good humor were compared by the people of Dayton with Bryan's solemnity, dignity, and show of virtue. Messenger boys who came to us would receive tips of quarters or half dollars; they got no tips from Bryan. Darrow would eat in a public restaurant, attend public affairs, even a high school dance; Bryan ate at home, attended funerals, and religious services. Darrow talked with individuals; Bryan preached to groups. The people came to prefer the human being to the messiah. They had been brought up on Bryan's ideas, but they liked Darrow's behavior" (Hays, 1942). Unlike Bryan, Hays believed that majoritarianism could be oppressive.

On July 21, Hays—who Potter claimed was "the greatest lawyer at the Dayton trial"—shocked everyone when he called Bryan as a witness (Potter, 1951). Hays later described the confrontation between Bryan and Darrow: "Perhaps never in the history of the world has a witness on cross-examination attempted rationally to defend beliefs not based on reasonable evidence." According to Hays, when Bryan admitted that the world was not necessarily created in six twenty-four-hour days, "Darrow demolished the fundamentalist case" (Hays, 1942). (Hays was apparently unaware that Bryan had made such an admission years earlier; see Mr. Bryan's rejoinder, 1923). Hays and co-counsel Neal did not get as much publicity as Darrow, but Scopes knew that without them, the speeches of Malone and Darrow "would never have shown so brilliantly."

In the trial's closing minutes, Hays offered to send Judge Raulston a copy of Darwin's *On the Origin of Species* and *The Descent of Man*. When Raulston replied, "Yes, yes" to Hays's offer, spectators in the courtroom laughed and applauded. Hays's offer was the last attorney's voice in the trial's transcript. Hays later claimed that the public "learned more about evolution through the Dayton exhibition than they could have in any other way" (Linder, 2004). When Bryan died five days after the trial, Hays noted, "Had this happened to Darrow, Tennessee, would have regarded it as a judgment of God. As it was, Bryan was gathered to the angels" (Scopes and Presley, 1967).

After the trial, Hays returned to New York, where he continued to fight for "the right of a man to be out of step with society." On April 20, 1931, Hays and Dudley Field Malone faced off in *The Trial of Napoleon Bonaparte*, a popular radio show sponsored by the National Dairy Products Corporation. The trial was broadcast on NBC radio stations and was decided by a jury that included U.S. Senators, university presidents, and Congresswoman Ruth Bryan Owen, the daughter of William Jennings Bryan. Hays was Bonaparte's chief prosecutor, and Malone was Bonaparte's defender (Hays and Malone, 1931). Bonaparte was exonerated by a vote of 8 to 4.

Like Darrow, Hays became rich representing the interests of power, privilege, and fame, including large corporations. Hays also served without fee in several notable civil-rights cases, including the Sweet segregation case in Detroit (1925), at which he again worked with Darrow; the *American Mercury* case involving H.L. Mencken about Boston's obscenity laws (1926); the Sacco-Vanzetti trial of two Italian anarchists in Boston who were convicted and executed for a murder they denied committing (1921–1927); and the Scottsboro case, in which eight black men were sentenced to death for attacking two white women (1931). Hays also defended striking coal-miners, people facing deportation, and religious groups (e.g., Jehovah's Witnesses), as well as people who refused to salute the American flag (Hays, 1942).

In 1951, Hays announced that Wisconsin Senator Joseph McCarthy was "the most dangerous man in the United States. [McCarthy] is more dangerous to freedom in the United States than all the Communists we have in this country ... he's dangerous because without evidence, he is smearing a lot of respected and highly decent people" (ACLU co-founder, 1951). Hays's books included *Let Freedom Ring* (1928, and revised in 1937), *Trial by Prejudice* (1937), and *Democracy Works* (1939). His colorful autobiography—*City Lawyer: The Autobiography of a Law Practice* (1942)—was Hays's last book. Until the end of his life, Hays lamented how "the religious views of the Middle Ages [could] persist in this age of railroads, steamboats, the telephone, the radio, the airplane, all the great mechanistic discoveries" (Kennedy, 1957).

The 73-year-old Hays died of a heart attack in New York City on December 14, 1954. He is buried in Beth-El Cemetery in Ridgewood, New York. In 1958, New York University founded its Arthur Garfield Hays Civil Liberties Program in his honor.

Sources

ACLU co-founder Arthur Garfield Hays on McCarthyism. 1951. Longines Chronoscope, December 7. C-SPAN. Accessed October 14, 2020, at https://www.c-span.org/video/?437422-1/longines-chronoscope-arthur-garfield-hays

Hays, Arthur Garfield. 1937. *Let Freedom Ring*. New York: Boni and Liveright, 27–28.

______. 1942. *City Lawyer: The Autobiography of a Law Practice*. New York: Simon & Schuster.

______, and Dudley Field Malone. 1931. *The Trial of Napoleon Bonaparte*. New York: Longacre Press.

Kennedy, Gail (Editor). 1957. *Evolution and Religion*. New York: D.C. Heath, 36.

Linder, Doug. 2004. Arthur Garfield Hays. Accessed October 18, 2020, at http://law2.umkc.edu/faculty/projects/ftrials/scopes/haysarth.htm.

Malone glad trial starts on Friday. 1925. *Chattanooga Times*, July 10, 2.

Mr. Bryan's rejoinder. 1923. *The Forum*, 70 (July–December), 1852–1853.

Potter, Charles Francis. 1951. *The Preacher and I: An Autobiography*. New York: Crown.

Scopes, John T., and James Presley. 1967. *Center of the Storm: Memoirs of John T. Scopes*. New York: Holt, Rinehart and Winston, 217.

Hicks, Herbert Eugene
1893–1946

Herbert Eugene Hicks was born on February 6, 1893, in Madisonville, Tennessee. His parents, Charles Wesley Hicks and Susan Hicks, were pioneering citizens of Monroe County, Tennessee. Herbert served in the Navy after World War I aboard the battleships *New Mexico* and *Pennsylvania*, and graduated from the University of Kentucky in 1920. He then began practicing law with his brother, Sue, in Dayton's Hicks & Hicks Law Firm.

In 1925, Herbert was serving as interim county attorney for Rhea County. He admitted that he and his brother, Sue, "think there is some evidence for [evolution], but until the theory is proven we see no reason for abandoning the Bible" (Anderson, 1925). When John Scopes was indicted, Herbert, his brother Sue, Wallace Haggard, and Rhea County School Superintendent Walter White were the first people associated with Scopes' prosecution; as trial instigator George Rappleyea admitted, "I couldn't find a worse pair of fundamentalists in the world than [the Hicks brothers], unless it would be Wallace Haggard, so I'll employ the three of you as special prosecutors" (Anderson, 1925). In early June, Herb, Sue, and Haggard met with Bryan in Nashville to plan the prosecution's strategy, as they did later that month in Atlanta. The ACLU agreed to pay the group's expenses, but they all declined the offer.

Herbert's most prominent role during the trial occurred on Thursday, July 16. After reminding Judge Raulston that "the issue of fact for the jury to determine is whether or not Prof. Scopes taught that man descended from the lower order of animals," Hicks argued that scientists' testimony should not be allowed because "we are not endeavoring to run here a teachers' institute; we do not want to make out of this a high school or college.... Why should these experts know anything more about the Bible than some of the jurors?" Herbert believed that Darrow and the defense "would attack the Bible ... [and] make our case for us. You know how juries are in this county" (Anderson, 1925). On the day of the verdict, Herbert bragged to his brother Ira (a fundamentalist preacher in New Jersey) about defeating the "agnostic" Darrow, the "atheist" Hays, and the "ostracized Catholic" Malone, but wondered if the prosecutors would get credit for their accomplishment from the biased media (Herbert Hicks to Ira Hicks, 1925). When Bryan died five days after the trial, Herb and Sue Hicks, Gordon McKenzie, and Wallace Haggard watched Bryan's body through the night.

After the trial, Herbert continued to work with his brother Sue in Dayton's Hicks & Hicks Law Firm until 1930, when he was paralyzed by a stroke. The 53-year-old Hicks died on April 27, 1946, in a Knoxville hospital. He is buried In Madisonville Cemetery in Madisonville, Tennessee.

Sources

Anderson, Paul Y. 1925. The drugstore argument that swelled to a national debate. *St. Louis Post-Dispatch*, May 31, 17, 20.

Hicks, Herbert, to Ira Hicks. 1925. July 22, in Judge Sue K. Hicks Papers, University of Tennessee Libraries, Knoxville, Tennessee.

Hicks, Sue Kerr 1895–1980

Sue Kerr Hicks was born in Madisonville, Tennessee, on December 12, 1895, as the youngest of nine children to Charles Wesley and Susan Hicks. When his mother died shortly after giving birth to him, his grieving father named him Sue to honor his late wife. Sue attended Hiwassee College and the University of Kentucky, after which he became an attorney and moved to Dayton, where he joined his older brother, Herbert, who was serving as acting Rhea County attorney. Sue and Herb formed Hicks & Hicks Law Firm, which was active throughout Rhea County. Another of Sue's brothers, Ira Evans Hicks (1875–1941), was a New Jersey Presbyterian minister who Sue consulted before and during the trial. More than a month before the trial began, Ira promised Sue and their brother Herbert that "the advertising this case will bring you will be worth untold money.... You will have no trouble making the evolutionists who take the stand look like a joke" (Farrell, 2011). Ira also warned Sue that getting facts about the trial to people would "be a difficult task," but Sue knew that "all we have to do is get the fact that Mr. Darrow is an atheist and does not believe in the Bible across to the jury, and his case is lost" (Stone, 1943).

When Bryan arrived in Dayton, he worked in the Hicks's office, during which time he told the Hicks brothers that they "would probably live to know whether or not evolution was true, but he would not." Throughout the trial, Hicks claimed that Bryan "would just walk up to strangers on the street and start talking with them," while Darrow "was shrewd and undoubtedly was a brilliant lawyer" who "never went out of his way to speak to anybody" (Glenn, 1975). According to most reports, Sue did "most of the heavy work in preparing for the trial" (Moutoux, 1925).

At the Scopes Trial, Sue and the other prosecutors were overshadowed by Bryan. Hicks admired how Judge Raulston handled the trial, later noting, "All manner of rabble [was] assembled at this trial, and at times the excitement of the crowd became almost a frenzy ... the attorneys for the defense did everything they could to provoke the Court and to get on the front pages of the newspapers ... so the situation was very hard to handle" (Sue K. Hicks, 2015). Hicks, who admitted that the "snobbish" Darrow's calling Bryan as a witness "was a shrewd move," recalled Bryan as "an outgoing man who lived up to his description as the Great Commoner during his stay in Dayton" (Glenn, 1975).

When Bryan died, Hicks and Bryan's nephew Bryan Jennings accompanied Bryan's body to Chattanooga. Hicks then took Jennings's advice and moved to Miami, Florida, where he worked with the Walsh, Beck, & Ellis law firm for four years. In 1930 he returned to live in Tennessee. When the movie version of *Inherit the Wind* was released in 1960, Hicks was so outraged by its portrayal of Bryan ("They had him gnawing chicken bones in the courtroom") that he considered buying television time to explain what really happened.

Sue was later immortalized in a song named "A Boy Named Sue," which was recorded by Johnny Cash. The song was written by Shel Silverstein, who got the idea for the song after hearing Hicks speak at a judicial conference in Gatlinburg, Tennessee (which is mentioned in the song's fourth verse). After the song became a hit, Cash sent Hicks two records and autographed photos inscribed with a paraphrased lyric from the song, "To Sue, how do you do?" In return, Hicks sent Cash an autographed photo of himself (Johnny Cash is indebted, 1970). Cash's "A Boy Named Sue" won a Grammy Award in 1970.

In the fall of 1934, Hicks ran (as "S.K. Hicks") for state representative with ads promising help and "relief instead of promises." Hicks defeated Walter White and served one term for Rhea County. While a legislator, Hicks—who claimed that the Butler Act had not affected anyone "except psychologically"—voted against repealing the Butler Act, reminding his colleagues that evolution remained an unproven theory that would make students question their faith. Hicks then moved to Madisonville, where he served as Criminal Court Judge for the Fourth Judicial District from 1936 to 1958, and thereafter as a senior reserve judge until 1972. Although he tried more than 800 murder cases, Hicks admitted, "The most publicity [I've received] has been from the name 'Sue' and the evolution trial" (Glenn, 1975). To commemorate the Scopes Trial's 50th anniversary, Hicks recalled that he liked Bryan "very much" and was a "little bit timid" around him, but respected the fact that Bryan helped the prosecution without telling the attorneys what to do. Hicks claimed that Darrow was a "brilliant lawyer" but a "ruthless kind of person," adding that although his interrogation of Bryan hurt the prosecution, "Bryan did as well as anyone could." Hicks believed that the Scopes Trial accomplished little except bringing ridicule to Tennessee (Sue K. Hicks interview, 1975).

The 84-year-old Hicks died on June 17, 1980, in a nursing home in Sweetwater, Tennessee. He is buried beside his wife, Reba Sue Bradley Hicks (1910–2000), in Hill Memorial Gardens in Madisonville, Tennessee.

Sources

Farrell, John A. 2011. *Clarence Darrow: Attorney for the Damned*. New York: Doubleday, 370.
Glenn, Juanita. 1975. Judge still recalls "Monkey Trial"—50 years later. *Knoxville Journal*, July 11, 17.
Johnny Cash is indebted to a judge named Sue. 1970. *New York Times*, July 12, 66.
Moutoux, John T. 1925. Young lawyers may achieve fame in Scopes case. *Knoxville News-Sentinel*, June 4, 6.
Stone, Irving. 1943. *Clarence Darrow for the Defense*. Garden City, NY: Doubleday, 443–444.
Sue K. Hicks ("A Boy Named Sue"). 2015. Accessed August 29, 2020, at https://www.geni.com/people/Sue-K-Hicks-A-Boy-Named-Sue/6000000003187074267.
Sue K. Hicks interview. 1975. University of Tennessee Library and Archives, ID #33817.

Hunter, George William 1863–1948

George William Hunter was born on June 29, 1863, in Mamaroneck, New York. After earning a degree in biology from Williams College (Massachusetts) in 1895, he spent a summer at Woods Hole (then called Woods Holl) Biological Station, and in 1918 earned a doctorate from New York University.

Some of Hunter's most productive years occurred when he chaired the biology department at his alma mater, DeWitt Clinton High School, a public boys school in Hell's Kitchen of New York City. There, he often worked with educators at Columbia University Teachers College, including famed geneticist Thomas Hunt Morgan, as he wrote a variety of biology textbooks. Hunter and his colleagues at DeWitt Clinton developed the modern secondary-school biology curriculum (Magat, 2006).

In 1914, Hunter published *A Civic Biology: Presented in Problems*, in which he praised Darwin for "the proofs of the theory on which we today base the progress of the world." Hunter, who dedicated only five of the book's 432 pages to evolution, described Darwin's theory as the "belief that simple forms of life on the earth slowly and gradually gave rise to those more complex.…" Like several other textbooks of the time, Hunter's book endorsed White supremacy and eugenics. It was a bestseller; the 3,274 copies sold in Tennessee covered 90 percent of the market. While substitute teaching in W.F. Ferguson's biology class in the spring of 1925, John Scopes used Hunter's textbook and, in the process, made the textbook a notorious book (Shapiro, 2008).

Although Hunter believed that evolution is compatible with the Bible, there is no evidence that Scopes' defender Clarence Darrow wanted or asked Hunter to testify at Scopes' trial. (Darrow opposed eugenics, which Hunter explicitly endorsed.) Instead, on the night before the trial started, Darrow sent a telegram asking biology textbook author Benjamin Gruenberg (1875–1965)[20] to come to Dayton, noting "the defense would be delighted to add you[r] authority to our position." Gruenberg initially accepted Darrow's invitation, but quickly reneged after being told by his publisher that testifying at the trial would "kill" sales of his book in the South. In Florida, "Gruenberg's *Biology* or the People of Florida—Which?" became "the opening gun in the assault on the state legislature" (Allen, 1926).

During the trial, Scopes' prosecutors focused more on Hunter's textbook than on Scopes. According to the prosecution, if Scopes used Hunter's textbook, then he taught evolution, and he therefore broke the law. This argument, rather than the validity of evolution, was what prosecutor Tom Stewart tried to focus on during the trial.

Hunter later worked at Pomona College and the Claremont Graduate School. He died on February 4, 1948, at age 74 in Claremont, California, and is buried in Hillside Cemetery in Redlands, California.

Sources

Allen, Harbor. 1926. The anti-evolution campaign in America. *Current History*, 24 (6), 893–897.
Magat, Richard. 2006. The forgotten roles of two New York City teachers in the epic Scopes Trial. *Science & Society*, 70 (4), 541–549.
Shapiro, Adam R. 2008. *Civic Biology* and the origin of the school antievolution movement. *Journal of the History of Biology*, 41 (3), 430.

Inherit the Wind

The Play

In 1951, American playwrights Jerome Lawrence (1915–2004) and Robert E. Lee (1918–1994) used a fictitious retelling of the Scopes Trial to defend "the right to think." Their cautionary, three-act play was *Inherit the Wind*, whose title comes from Proverbs 11:29: "He that troubleth his own house shall inherit the wind...." Lee believed that "today, more than at any time in 100 years, basic intellectual freedoms have been challenged.... We're not preaching. We just hope when people get home after seeing the show, they'll suddenly say to themselves, 'Hey, that had some relevance to life today.'"

Lawrence and Lee admitted that their play was not about the Scopes Trial; it was about the intolerance and zealotry of McCarthyism. Although Lawrence and Lee interviewed several of the Scopes Trial's participants (e.g., Arthur Garfield Hays), the playwrights warned readers that "*Inherit the Wind* is not history. The events which took place in Dayton, Tennessee, during the scorching July of 1925 are clearly the genesis of this play. It has, however, an exodus entirely its own.... *Inherit the Wind* does not pretend to be journalism. It is theatre. It is not 1925.... It might have been yesterday. It could be tomorrow" (Lawrence and Lee, 1955). Instead of being set in Dayton, Tennessee, in July 1925, *Inherit the Wind* happens in Hillsboro, "a small town" in an unnamed state during a summer "not too long ago." Several of the play's most prominent characters have names that sound similar to (and have the same number of syllables as) their counterparts from the Scopes Trial. For example, Henry Drummond is patterned after Clarence Darrow and is the mouthpiece for science and reason; E.K. Hornbeck is a cynical reporter for the fictional *Baltimore Herald* who is patterned after H.L. Mencken; Bertram Cates is a Hillsboro teacher patterned after John Scopes; and Matthew Harrison Brady is patterned after William Jennings Bryan. Lawrence and Lee also created several characters that had no counterparts in Scopes' actual trial, including Rachel Brown, Cates's girlfriend who struggles to see whose view is correct (eventually siding with Cates), and her father, the stern-faced the Rev. Jeremiah Brown, the fanatical leader of Hillsboro's fundamentalists. The Reverend Brown, during an outdoor, torch-lit sermon, excites his followers by asking, "Do we curse the man who denies the Word?"; "Do we cast out this sinner in our midst?"; and "Do we call down hellfire on the man who has sinned against the Word?," after each of which the crowd screams "Yes!"

Lawrence and Lee also invented several events; for example, the citizens of Dayton (unlike those of Hillsboro) did not attack or despise Scopes, and Scopes (unlike Cates) was a willing participant in his trial, was not a social outcast, was not arrested in his classroom, did not have a girlfriend during the trial, and was never jailed. Bryan (unlike Brady) was not a vain, narrow-minded, untrustworthy glutton, nor did he die after having a fit in the courtroom. By the end of the play, Drummond is the victorious, tolerant hero, and Brady—who early in the play is a voice of reason when protecting Rachel from her rabid father—becomes an intolerant, irrational loser. The public interpreted *Inherit the Wind* as a documentary about the Scopes Trial, but it tells viewers more about the 1950s (i.e., when it was written) than 1925 (when the Scopes Trial occurred).

Producing *Inherit the Wind*

After finishing *Inherit the Wind*, Lawrence and Lee shopped their play to Broadway, but all eight producers they approached turned them down. They finally found a suitor in Dallas, Texas, where producer Margo Jones (1911–1955) read the script and told them, "You have certainly given me life through this script." When Jones staged *Inherit the Wind* at her Theatre '55, Lawrence and Lee told reporters, "Margo saw in *Inherit the Wind* the conflict and the drama which not one other person in the American theater saw" (Cattarulla, 2005). Little did Jones know that she was launching *Inherit the Wind* on its way to becoming a classic of American theater.

Jones premiered *Inherit the Wind* on January 10, 1955, at her Theatre '55. Her 22-person play, staffed with local actors, was well received. For example, critic John Rosenfield of the *Dallas Morning News* described it as a "new play of power, humanity and universal truth" that would be popular in New York because it depicted "the South that New Yorkers love: the South of tapeworms, ignorance, and bigotry" (Rosenfield, 1955). *Inherit the Wind* attracted record crowds to Jones's theater. From 1,500 miles away, even the *New York Times* noticed that a "play based on the Scopes Trial has premiered in Dallas" (*Inherit the Wind* opens, 1955).

After some legal maneuvers, Broadway producer Herman Shumlin (1898–1979)—who had rejected the play earlier—obtained the rights to *Inherit the Wind*. His 60-person play premiered on April 21, 1955, at Broadway's National Theater and featured Ed Begley (playing Brady), Tony Randall (in his first major stage role, playing Hornbeck), and an out-of-retirement Paul Muni (playing Drummond). Shumlin, whose advertisements claimed the play was "based on the thrillingly penetrating Scopes Trial," invited John Scopes to the play, but Scopes turned him down because he was "too busy and not interested enough to fly to New York to see it" (Scopes and Presley, 1967). Most of the play's reviews were glowing, and *Inherit the Wind* became the talk of the town. When *Inherit the Wind* closed on June 22, 1957, its 806 performances made it Broadway's most successful and longest-running drama. As Randall later noted, "I thought the only people who believed the Bible literally were kooks." He went on: "I did not think the play would be taken seriously; the fundamentalists had become a lunatic fringe.... You think this is over? It's never over" (Mandell, 1996).

Inherit the Wind has been translated into more than 30 languages and has been revived repeatedly on Broadway and elsewhere. Lawrence and Lee's play, far more than what actually happened in Dayton, is now the basis for most people's "knowledge" of the

Scopes Trial. Indeed, *Encyclopædia Britannica* included its first entry for "The Scopes Trial" in 1957 after *Inherit the Wind* appeared on Broadway. That entry, which referred to *Inherit the Wind*, noted that the trial seems to have been about "a religious movement, Fundamentalism, [rather than about] an individual" (*Encyclopædia Britannica*, 1957). *Inherit the Wind*, which remains relevant to many of the tensions that animate modern society, continues to be performed often throughout the world.

The Movies

In March 1956, American filmmaker Stanley Kramer (1913–2001) bought the screen rights to the play *Inherit the Wind* for "something like $200,000," after which he hired screenwriters Harold Jacob Smith (1912–1970) and Nedrick Young (1914–1968)[21] to adapt the play to the silver screen (Kramer and Coffey, 1997). Kramer's 128-minute movie, which starred renowned actors Spencer Tracy (as Drummond) and Frederic March (as Brady), was promoted as being "All About the 'Monkey Trial,' Rocked America!" After showings at the Berlin International Film Festival and in London, *Inherit the Wind* had its North American premiere on July 21, 1960, at Dayton's Scopes Trial Day.

The movie differs significantly from the play. For example, in the movie, Cates (i.e., the Scopes character) is arrested in his science classroom by town leaders for teaching evolution, and Brady's wife (i.e., the wife of the Bryan character) is more prominent than in the play. The ACLU, which helped instigate the trial, is not in the movie, nor is Darrow's co-counsel (e.g., Dudley Field Malone, who gave the trial's most stirring speech). Instead of being portrayed as a coach and a substitute biology-teacher who knew little about evolution, Cates is a martyr dedicated to teaching evolution. A fictional romance between Cates and Rachel is dramatized, as Scopes himself suggested, for "the balcony set" (Scopes and Presley, 1967). The movie also includes several historically accurate events, including Bryan's advocacy of majoritarianism, Bryan's campaigns for president, Scopes' not testifying at his trial, and Darrow's citation for contempt of court (Moore, 1999). As in the play, the fanatical Reverend Brown represents fundamentalism, and Brady dies a reactionary.

Although Kramer's movie was boycotted by the American Legion and elicited protests "at every theater that showed the picture," it was nominated for a variety of awards, both in the United States and abroad (Kramer and Coffey, 1997). *Time* magazine dismissed Kramer's movie as "a sluggish, confused manipulation of ideas and players," and *The New Yorker* complained, "History has not been increased but almost fatally diminished" (Gibbs, 1955). Other critics praised it as "sensational, and I mean sensational!" (*New York Daily News*) and "one of the most brilliant and engrossing displays of acting ever witnessed on the screen" (*New York Times*; United Artists, 1960). Late in 1960, *Inherit the Wind* became the first in-flight movie when Trans-World Airlines (TWA) showed it aboard a flight from New York to Los Angeles. Despite its publicity and mostly positive reviews, Kramer's *Inherit the Wind* lost money (Kramer and Coffey, 1997).

When she saw *Inherit the Wind* at its premiere in Dayton, Bryan's granddaughter Helen Rudd Brown (1920–2015)—who admittedly didn't "know much about my grandfather"—enjoyed the "extremely good film" (Ogden, 1960). However, Scopes' prosecutor Sue Hicks was outraged by the movie's treatment of Bryan, as were other residents of

On July 21, 1960, Stanley Kramer's movie *Inherit the Wind*—claiming to be "All About the Fabulous 'Monkey Trial' That Rocked America!"—had its North American premiere at the Dayton Drive-In Theater. Today, this fictitious movie, rather than what actually happened in Dayton, is the basis for most people's "knowledge" of the Scopes Trial.

Dayton. For example, Eloise Reed, who attended the trial as a 13-year-old, recalled that at a reception in Dayton, "Kramer came over [to me] and sat down beside me, and said, 'What did you think of the film?' and I said, 'I didn't like it. I thought it was awful.... [Dayton and the Scopes Trial weren't] like that at all'" (American Experience, 2002). To John Scopes, however, "the film captured the emotions in the battle of words between Bryan and Darrow" (Scopes and Presley, 1967).

Other Interpretations

In late 1965, NBC broadcast a remake of *Inherit the Wind* that was less humorous and more true to the play than was Kramer's movie. In 1988, NBC presented another remake that was less melodramatic than Kramer's original film. Its stars included Kirk Douglas, who played Brady. Whereas Frederic March had modeled his 1960 portrayal of Brady on William Jennings Bryan (e.g., with a balding wig and stomach padding), Douglas modeled his version of Brady on contemporary televangelists Oral Roberts, Jimmy Swaggart, and Pat Robertson, the latter of whom, like Bryan, had ran unsuccessfully for president. In 1999, the Showtime Network produced yet another remake of *Inherit the Wind* that was advertised as a "dramatization of the 1925 Scopes 'monkey trial.'" This remake, like the others, lacked the impact of Kramer's original.

In 2007, Bryan College responded to *Inherit the Wind* with *The Scopes Trial: Inherit the Truth*, a 112-minute, black-and-white movie that was produced using local actors in Dayton. This movie, unlike all previous ones, focused on courtroom drama, for which it relied on the trial transcript. The movie ends when Judge Raulston adjourns the trial. *Inherit the Truth* follows the trial transcript closer than any other treatment of the Scopes Trial.

The Scopes Trial and *Inherit the Wind*

Just as Charles Dickens's popular *A Tale of Two Cities* (1859) imprinted on the public an inaccurate view of the French Revolution, so too did *Inherit the Wind* become the public's inaccurate "knowledge" of the Scopes Trial. Despite the disclaimer that *Inherit the Wind* was not accurate or historical, the public considered the movies and play to be accurate *and* historical as it substituted the movies' characters and events for what actually happened in Dayton. No one appreciated this more than John Thomas "Jack" Scopes III, who often told audiences, "The thing I'm most frequently asked is if my grandfather ever married the preacher's daughter" (Hayes, 1992). Teachers even began using the fictitious movie to teach about the actual trial (National Center for History in Schools, 1994; Larson, Depew, and Isetti, 2008).

Sources

American Experience. 2002. Monkey Trial. WGBH Educational Foundation and Nebraskans for Public Television, Inc., 1:16:55–1:17:18.

Cattarulla, Kay. 2005. The Texas Tornado. *Legacies, A History Journal for Dallas and North Central Texas*, 16 (2), 39–47.

Encyclopædia Britannica. 1957. The Scopes Trial. Chicago: Encyclopedia Britannica, Vol. 20, 134.

Gibbs, Wolcott. 1955. Mixed bag. *New Yorker*, 31 (11), 67.

Hayes, Ron. 1992. Inheriting the wind. *Palm Beach (FL) Post*, May 29, 1D, 7D.

Inherit the Wind opens—play based on Scopes Trial has premiere in Dallas. 1955. *New York Times*, January 12.
Jones, Margo, to Lawrence and Lee, November 10, 1954. Margo Jones Collection, Dallas, Texas.
Kramer, Stanley, with Thomas M. Coffey. 1997. *A Mad, Mad, Mad, Mad World: A Life in Hollywood*. New York: Harcourt Brace, 172.
Larson, Edward J., David Depew, and Ronald Isetti. 2008. Inheriting Inherit the Wind: Debating the play as a teaching tool. *Evolution Education and Outreach*, 1, 150–157.
Lawrence, Jerome, and Robert E. Lee. 1955. *Inherit the Wind*. New York: Random House.
Mandell, Jonathan. 1996. Inherit the controversy. *Newsday*, March 17, 10.
Moore, Randy. 1999. Creationism in the United States. VII. The lingering impact of *Inherit the Wind*. *American Biology Teacher*, 61 (4), 246–249.
National Center for History in Schools. 1994. *National Standards for United States History: Exploring the American Experience*. Los Angeles: National Center for History in Schools, 180.
Ogden, Warner. 1960. Scopes challenges wording of marker. *Knoxville News-Sentinel*, July 24, A8.
Rosenfield, John. 1955. *Inherit the Wind*: Arena stage's finest hours. *Dallas Morning News*, January 11, 12.
Scopes, John T., and James Presley. 1967. *Center of the Storm: Memoirs of John T. Scopes*. New York: Holt, Rinehart and Winston, 266.
United Artists. 1960. *Pressbook*. Undated booklet in archives of Rhea County Historical Society and Genealogical Society, Dayton, Tennessee.

Jones, Charles R.
1875–1967

Charles R. Jones was born on September 9, 1875. At the time of the Scopes Trial, the Reverend Jones—known in Dayton as "Parson Charlie"—pastored Dayton's First United Methodist Church (South) on Market Street. There, on the first Sunday of the trial, William Jennings Bryan delivered the morning sermon, before which Jones introduced him as an "ambassador for Christ." With Judge John Raulston and his family sitting in the front row, Bryan—who had not yet spoken in the courtroom—preached (and promoted Florida real estate) to an overflow crowd for 70 minutes. Nine days later, on the last day of trial, Jones delivered the trial's benediction.

On Sunday, July 26, Bryan made a surprise appearance at Jones's church, where Jones called on him to lead a prayer. This was Bryan's last public appearance. Bryan then went to the home of F.R. Rogers, just a few blocks from Jones's church where Bryan and his entourage were staying. When Bryan died there that day during an afternoon nap, Jones went to the Rogers's home to console Bryan's widow, Mary, who was sitting in her wheelchair in the front yard. When Jones told Mary that "the workman dies, but the work is carried on," Mary "looked up at me and said, 'But who will carry on?'" (Moore, n.d.).

On July 28, Jones officiated Bryan's memorial service on the front lawn of the Rogers's home. Jones praised the "young" lawyers who stood by [Bryan's] side in defense of the greatest Book that has ever been looked upon. May the mantle of William Jennings Bryan have fallen upon them, and may they, together with others, take up the great work where he has laid it down and carry it on to victory.... We pray thy blessing upon Mrs. Bryan, who is left to mourn the loss of one of the greatest men who ever lived" (Preacher prays Dayton lawyer may take mantle, 1925). Several days later, Jones led another memorial service for Bryan at the Baptist Tabernacle in Chattanooga.

After the Scopes Trial, Jones continued to preach in Dayton and elsewhere. He died on February 25, 1967, in Dayton, where he is buried in Dayton's Buttram Cemetery.

Sources

Moore, Hugh. n.d. Dayton minister recalls conducting funeral of Bryan. Undated newspaper article, Rhea County Historical Society and Genealogical Society Museum.
Preacher prays Dayton lawyer may take mantle. 1925. *Chattanooga Daily Times*, July 29, 1.

Jones, Margaret Virginia "Margo" 1911–1955

Margaret Virginia "Margo" Jones was born on December 12, 1911, in Livingston, Texas. After graduating from Livingston High School at age 15, she earned an undergraduate degree (in 1932) and a master's degree (in 1933) from what is now Texas Woman's University. Three years later, Jones founded the Houston Community Players, and from 1942 to 1946 she taught drama at the University of Texas in Austin. In November 1947, the energetic Jones—often called "The Texas Tornado"—opened Dallas's first nonprofit resident theater, which she named Theatre '47. In subsequent years, Jones changed the digits in her theater's name to emphasize her commitment to producing new plays (e.g., in 1948, Theatre '47 became Theatre '48). Throughout her career, Jones staged 85 plays, 57 of which were world premieres. Theatre '47 was the nation's first professional arena theater-in-the-round. Jones hoped that if such theaters were successful, playwrights would not need Broadway.

On January 10, 1955, Jones staged the world premiere of Jerome Lawrence and Robert Edwin Lee's *Inherit the Wind* at her Theatre '55. The three-act play, which had been rejected by eight Broadway producers, featured 22 actors, most of whom were from the Dallas area. Jones's well-received production marked the beginning of *Inherit the Wind* becoming a classic of American theater and a major influence on the public's perception of the Scopes Trial. Today, *Inherit the Wind* remains one of the most-performed plays in the world.

After noting the success of *Inherit the Wind* in Jones's theater, Broadway producer Herman Shumlin—who had earlier rejected Lawrence and Lee's script—obtained the rights to produce the play on Broadway. After suing to stop Jones from extending *Inherit the Wind*'s run in Texas, Shumlin—"in association with Margo Jones"—premiered *Inherit the Wind* at Broadway's National Theatre on April 21, 1955. *Inherit the Wind* ran for a then-record 806 performances.

On July 17, 1955, while *Inherit the Wind* was attracting sellout crowds on Broadway, Jones spilled some red paint on her home carpet. Workers using carbon tetrachloride to clean the spill let large amounts of the solvent sink into the carpet. When Jones fell asleep on an adjacent couch, she was poisoned by the fumes. Jones—a pioneer in regional theater—died of uremic poisoning on July 24, 1955, in Dallas, Texas, and is buried in Livingston's Forest Hills Cemetery. Jones is commemorated in Livingston by a historical plaque recognizing her as a "world-famed genius of drama." Jones, whose theater closed in 1959, is honored by the Margo Jones Theatre at Southern Methodist University and the Margo Jones Award, which was established in 1961 by Lawrence and Lee.

Sources

Catarulla, Kay. 2005. The Texas Tornado. *Legacies, A History Journal for Dallas and North Central Texas*, 16 (2), 39–47.

Jones, Margo, to Lawrence and Lee, November 19, 1954. Margo Jones Collection, Dallas Public Library, Dallas, Texas.
Rosenfield, John. 1955. *Inherit the Wind*: Arena stage's finest hours. *Dallas Morning News*, January 11, 12.

Jury

When the Scopes Trial began, Scopes' defender Clarence Darrow admitted that he did not expect to find anyone in Dayton—a juror or otherwise—who accepted evolution. During jury selection, eight veniremen—including a preacher who was applauded by courtroom spectators when he announced that he had "preached against [evolution], of course"—were excused.

All of the jurors in the Scopes Trial were from Rhea County, and virtually all of them were farmers: W.F. Roberson (no church affiliation, farmer), J.W. Dagley (Methodist, farmer), James W. Riley (Baptist, farmer), W.G. Taylor (Southern Methodist, farmer), R.L. Gentry (Baptist, farmer and teacher), William "Billy" Smith (Baptist, farmer), Jesse R. Goodrich (Disciple of Christ, shipping clerk), John H. Bowman (Methodist Episcopal, farmer and cabinet maker), William G. "Bill" Day (Baptist, farmer), R.L. West (Baptist, farmer and carpenter), and John S. Wright (Baptist, farmer). The jury's foreman and "the outstanding figure among the jury men" was Jack R. Thompson, a Methodist and former U.S. Marshall who was well-known throughout eastern Tennessee ("Captain Jack" a courtly juror, 1925). Although several reporters noted Malone's fancy clothes during the trial, Darrow praised Thompson as "the best looking and most carefully dressed ... of anyone else in the room" (Darrow, 1932).

Judge Raulston admitted that the jury "had no idea what [the trial] was all about when the trial started," but John Scopes' father described the group as "one hell of a jury" (Farmers will try teacher, 1925). At least one juror was there "to get a good seat for the trial," but the jury was in the courtroom for less than four hours during the entire trial, prompting one writer on the last day of the trial to report,"It came as something of a surprise to followers of the trial today to discover that there still was a jury" (Small, 1925). On July 21, 1925, the jury deliberated less than 10 minutes before finding Scopes guilty. Each juror was paid $2 per day for their work. John Wright, the last surviving juror, died on March 5, 1966.

Sources

"Captain Jack" a courtly juror. 1925. *Chattanooga Daily Times*, July 16, 13.
Darrow, Clarence. 1932. *The Story of My Life*. New York: Charles Scribner's Sons, 269.
Farmers will try teacher. 1925. *New York Times*, July 11, 1.
Small, Robert T. 1925. Scopes jurymen kept in dark. *Chattanooga Daily Times*, July 21, 1–2.

Kenyon, Nellie
1898–1982

Nellie Kenyon was born on November 28, 1898, in Lexington, Kentucky. While attending high school and college, she worked on Saturdays, holidays, and during vacations as a reporter for the *Chattanooga News*, where her mother edited the "Society Page." Kenyon, who began her career when few journalists were women, was the only female reporter at the Scopes Trial.

In 1970, pioneering reporter Nellie Kenyon (center)—who had covered the Scopes Trial in 1925—was reunited with John Scopes in Nashville. To the left is Mildred Scopes, John's wife (RCHGS).

By the time Kenyon came to Dayton in July 1925 to cover the trial, she was already listed in the national *Who's Who in Journalism*. Kenyon arrived in Dayton two weeks before the trial, during which time she uncovered details about the trial's origins (Kenyon, 1925). When she saw Judge Raulston and complained about the "dirty" courtroom, it was cleaned and painted. Kenyon, who was issued the trial's first press-pass, attended virtually every session in the hot, steamy courtroom. Her dogged pursuit of stories at the Scopes Trial exemplified a trait that defined her career; "If there was a Pulitzer Prize for persistence," Kenyon often said, "I'd get it." Teamsters' Union leader Jimmy Hoffa agreed, praising Kenyon as "the most persistent woman I have ever known" (Daughtrey, 1966).

During the first weekend of the trial, Kenyon accompanied journalist H.L. Mencken to a meeting of Holy Rollers. Although other reporters fawned over Mencken's celebrity, Kenyon had little else to do with him, claiming later that he "wrote with his tongue in his cheek and didn't mean half" of what he wrote (Adamson, 1975). In 1935 —that is, the tenth anniversary of the Scopes Trial—Kenyon urged Dayton to erect a memorial to Rappleyea for starting the Scopes Trial, but her suggestion was ignored.

In 1932, Kenyon snuck into a train car carrying gangster "Scarface" Al Capone, who was on his way to serve an 11-year sentence in federal prison in Atlanta. Capone,

America's number one gangster, appreciated Kenyon's effort to meet him and gave Kenyon an interview. When Kenyon asked Capone who was going to handle his business while he was in prison, Capone famously answered, "You are, cutie." In 1938, Kenyon—who received numerous death threats—was credited with solving the robbery of Chattanooga's Highland Park Bank, and in 1950 she exposed a nationwide baby-selling ring based in western Tennessee whose clients included Hollywood stars. Kenyon also reported on visits by Warren Harding and Franklin Roosevelt, train wrecks, and meetings of Chattanooga's Ku Klux Klan. Kenyon earned many awards for stories less sensational than those that attracted so much publicity.

Kenyon—a pioneering reporter—retired in April 1970 and, on January 8, 1974, was honored at "Nellie Kenyon Day" sponsored by University of Tennessee. There, Kenyon admitted that she was "a rather timid soul—except when it comes to getting stories" (Welsh, 1974).

After a long illness, the 83-year-old Kenyon—a often-honored member of the Tennessee Journalism Hall of Fame—died on February 13, 1982, in Nashville's Belmont Health Care Center. She was buried at Forest Hills Cemetery in Chattanooga.

Sources

Adamson, June. 1975. Nellie Kenyon and the Scopes "Monkey Trial." 1975. *Journalism History*, 2 (3), 88–97.

Daughtrey, Larry. 1966. Why you don't hold out on Nellie Kenyon. *The Nashville Tennessean Magazine*, January 2, 6.

Kenyon, Nellie. 1925. How Scopes case really started is told for the first time. *Chattanooga Times*, June 29, 1–2.

Nellie Kenyon, who began career on "Chattanooga News," dies at 82. 1982. *Chattanooga Times*, February 15, A6.

Welsh, Pat. 1974. Friends to honor Nellie Kenyon at UT. *The Nashville Tennessean*, January 6, 6A.

Kramer, Stanley Earl
1913–2001

Stanley Earl Kramer was born on September 29, 1913, in Manhattan, New York. He grew up in Hell's Kitchen and, after graduating from DeWitt Clinton High School at age 15, earned a degree from New York University in 1933 at age 19. He then worked as a film editor and researcher before entering the military in 1942. During World War II, Kramer produced training films for the Army Signal Corps. After the war, when he learned that there were few jobs in Hollywood, Kramer founded Screen Plays, an independent production company. By 1950, he had renamed his company the Stanley Kramer Company, which produced Marlon Brando's screen debut, *The Men* (1950).

In the late 1950s, after buying the stage rights for the play *Inherit the Wind* for "something like $200,000," Kramer visited Dayton "to determine how much the town had changed" (Kramer, 1997). He then hired screenwriters Harold Smith (1912–1970) and Nedrick Young (1914–1968) to adapt the play to the silver screen. Filming, which began in October 1959, lasted 11 weeks. The outdoor scenes used more than 1,000 extras, and Kramer filmed the courtroom scenes over three weeks in front of 300 spectators (American Film Institute Catalog, 2019). *Inherit the Wind* was the first of four films in which Kramer worked with actor Spencer Tracy.

Inherit the Wind had its North American premiere at Scopes Trial Day in Dayton,

In 1960, John Scopes (left) met Stanley Kramer when Scopes was in Los Angeles promoting *Inherit the Wind*. Kramer points to Tennessee, where Scopes' famous trial had occurred 35 years earlier (RCHGS).

Tennessee, on July 21, 1960. After the showing, John Scopes told reporters that he "enjoyed the movie" and that the movie, despite its historical inaccuracies, "captured the emotions in the battle of words between Bryan and Darrow" (Scopes and Presley, 1967).

During production of *Inherit the Wind*, United Artists had reminded Kramer that they were "in business to make money." Kramer later admitted that their reminder was "a fact I seemed to have forgotten.... I might as well [have asked] them to back a golf course in Antarctica as a movie about the theory of evolution and a trial without a corpse.... I suppose I should have listened to them and dropped the idea, but I could not" (Kramer, 1997). Unlike Broadway's stage play of *Inherit the Wind*, Kramer's movie of *Inherit the Wind* lost money.

Kramer's movies garnered 16 Academy Awards and 80 nominations. Several of them, including *Inherit the Wind*, addressed topics that other producers avoided, including Nazi fascism (*Judgment at Nuremberg* in 1961), nuclear war (*On the Beach* in 1959), and racism (*Guess Who's Coming to Dinner* in 1967). Kramer's most commercially successful movie was the comedy *It's a Mad, Mad, Mad, Mad World* (1963). Although Kramer was known for his "message films," he insisted, "I'm not interested in changing anyone's mind, just in telling a story" (Spoto, 1978).

After directing his last film—the little-seen *The Runner Stumbles* (1979)—Kramer moved to Seattle, where he taught classes, wrote a newspaper column (for the *Seattle Times*), and hosted a weekly show about movies. In his biography, Kramer noted, "I've enjoyed a good career in the movie industry, yet not as good as my ambitions had led me to hope. My pictures were never, ever good enough to approach the dream" (Kramer and Coffey, 1997).

Kramer died from complications from pneumonia on February 19, 2001, at age 87 in the Motion Picture Home in Woodland Hills, California. Kramer's cremated remains were given to his family. On March 28, 1960, Kramer was among the first people to be honored with a star on the Hollywood Walk of Fame. In 2002, the Producers Guild of America honored Kramer by creating the Stanley Kramer Award, which is given annually to films or filmmakers who "dramatically illuminate provocative social issues."

Sources

American Film Institute Catalog. 2019. *Inherit the Wind*. Accessed September 10, 2020, at https://catalog.afi.com/Catalog/moviedetails/53192.

Kramer, Stanley, and Thomas M. Coffey. 1997. *A Mad, Mad, Mad, Mad World: A Life in Hollywood*. New York: Harcourt Brace.

Scopes, John T., and James Presley. 1967. *Center of the Storm: Memoirs of John T. Scopes*. New York: Holt, Rinehart and Winston, 270.

Spoto, Donald. 1978. *Stanley Kramer: Film Maker*. New York: Putnam, 18.

Ku Klux Klan
Est. 1865

The Ku Klux Klan (KKK) is an American White supremacist group having three distinctive eras or waves of activity in the United States: 1865–1871, 1915–1944, and 1946–present. The second wave of Klan activity was organized by Protestant nativists who expanded the post–Civil War's anti–Black sentiments to include immigrants, other faiths (especially Catholic and Jewish), and organized labor. During the height of the second-generation KKK, membership increased to over four million citizens nationwide and held considerable political influence, backing candidates at every level of government and becoming one of the largest social movements of the first half of the twentieth century. Nationwide marketing campaigns spurred by post-war nativist sentiment and the popular 1915 film *Birth of a Nation* made the Klan's now-iconic symbols—for example, Klan parades, white robes, and burning crosses—part of the Klan's message. Although the KKK had been founded in Tennessee and was active there throughout the 1920s, there was not a chapter in Dayton at the time of the Scopes Trial. However, Fay-Cooper Cole—a scientist who came to Dayton to testify for the defense—later claimed that policemen from Chattanooga were at the trial to protect him and other witnesses from the Klan (Cole, 1959).

In the 1920s, the KKK was part of the fundamentalist movement that helped produce the Scopes Trial. The Klan considered fundamentalists to be kindred spirits, for the Klan's emphasis on conservative Christianity and "100% American" patriotism meshed well with fundamentalists' ideals and their campaign to ban the teaching of evolution in public schools. When Oklahoma passed the nation's first antievolution law in 1922,

Governor John Walton acknowledged that "but for the influence of the KKK, I doubt if it could have carried either house" (Granados, 2012).

A few fundamentalist leaders sometimes disowned the Klan, but the Klan did not disown them. Most fundamentalist leaders were not members of the Klan, but they repeatedly accepted the Klan's support and money (Newton, 2014; Klansmen hand $50, 1922). Texas firebrand Frank Norris often hosted Klan meetings at his church, and at some of Billy Sunday's revivals, Klansmen were "guests of honor." The Klan's opposition to the teaching of evolution is noted in the movie *O Brother, Where Art Thou?* (2000), in which a Klan leader promotes "old time religion" while urging his fellow Klansmen to oppose "all those smart-ass folks [who] say we done descended from monkeys."

At the Democratic National Convention in June 1924, William Jennings Bryan—knowing that the Klan was popular in places that supported his campaign against evolution—opposed Democrats' plan to denounce the Klan.[22] Bryan believed that condemning the Klan by name would disrupt party unity ("Christians, stop fighting!"), but others believed not doing so endorsed the Klan's bigotry. When he addressed the convention, Bryan claimed that it required more courage to fight the Republican Party than it did to oppose the KKK. The resolution to condemn the Klan failed by one vote, after which Bryan claimed that the victory "must have been the Lord's work. I would not have been willing to risk so small a margin" (Maxwell, 2014; Granados, 2012).

Bryan was never a member of the Klan and he opposed much of what the Klan did. Nevertheless, the Klan claimed him as one of their own. When Bryan died in Dayton five days after the Scopes Trial, Klan chapters were named in his honor, and several chapters held memorial services that included burning crosses inscribed "In memory of William Jennings Bryan, the greatest Klansman of our time, this cross is burned; he stood at Armageddon and battled for the Lord" (Smith, 1969). Invoking Lincoln's "House Divided" speech from 1858, the Klan's Grand Dragon of Ohio announced, "we will take up the torch as it fell from [Bryan's] hand" because "America cannot remain half Christian and half agnostic" (Gordon, 2017). On August 9, 1925, after a massive rally of more than 30,000 Klansmen in Washington, D.C., the Klan laid wreaths atop Bryan's grave in Arlington National Cemetery. A year later, the Knox County (Tennessee) chapter of the Klan praised the William Jennings Bryan Memorial Association, adding, "you will find each and every Klansman ready to help … [we] know and realize what William Jennings Bryan stood for" (Maxwell, 2014).

Sources

Bryan, William Jennings. 1923. Bryan says North would act as South on Negro question. *New York Times*, March 18, Special Section, XXI.

_______, and Mary Baird Bryan. 1925. *The Memoirs of William Jennings Bryan*. Philadelphia: John C. Winston, 330.

Cole, Fay Cooper. 1959. A witness at the Scopes Trial. *Scientific American*, 200 (1), 120–131.

Gordon, Linda. 2017. *The Second Coming of the KKK*. New York: Liveright, 51.

Granados, Luis. 2012. *Damned Good Company: Twenty Rebels Who Bucked the God Experts*. Washington, D.C.: Humanist Press, 269.

Klansmen hand $50 to minister. 1922. *Cincinnati Enquirer*, May 12, 3.

Maxwell, Angie. 2014. *The Indicted South: Public Criticism, Southern Inferiority, and the Politics of Whiteness*. Chapel Hill: University of North Carolina Press.

Newton, Michael. 2014. *White Robes and Burning Crosses: A History of the Ku Klux Klan from 1866*. Jefferson, NC: McFarland.

O Brother, Where Art Thou? 2000. Touchstone Pictures, 1:16:12–39.

Smith, Willard H. 1969. William Jennings Bryan and racism. *Journal of Negro History*, 54 (2), 127–149.
Sutton, Matthew Avery. 2007. *Aimee Semple McPherson and the Resurrection of Christian America*. Cambridge: Harvard University Press, 115–116, 133.

Lawrence, Jerome | Lee, Robert Edwin
1915–2004 | 1918–1994

Jerome Lawrence Schwartz was born in Cleveland, Ohio, on July 14, 1915, to printer Samuel Schwartz and poet Sarah Rogen Schwartz. He enjoyed reading and often hitchhiked to New York City to see plays. After earning a B.A. from the Ohio State University in 1937 and doing graduate work at UCLA, Lawrence worked as a newspaper editor and reporter, and in 1939 he began working as a writer at Columbia Broadcasting System (CBS) in Los Angeles and New York City. Robert Edwin Lee, who grew up about 30 miles from Cleveland in Elyria, Ohio, was born on October 15, 1918, to engineer Claire Melvin Lee and teacher Elvira Taft Lee. Lee attended Ohio Wesleyan University, after which he moved to New York City.

In 1942, Lawrence met Lee in New York City and they began a lifelong collaboration that produced 39 plays and 14 Broadway productions. They joined the U.S. Army in the early 1940s, at which time they helped found the Armed Forces Radio Service, which informed and entertained troops overseas and stateside. After World War II, Lawrence and Lee reentered civilian life and continued their work together on plays, musicals, and scripts for television and radio. Many of their thought-provoking plays—for example, *The Gang's All* Here (1959, about government corruption) and *The Night Thoreau Spent in Jail* (1970, a study of Thoreau's resistance to war)—focused on the dignity of the human mind and the fight against censorship.

Among Lawrence and Lee's most famous works is *Inherit the Wind*, which used a fictionalized version of the Scopes Trial to examine McCarthyism and intellectual freedom. Lawrence and Lee had no interest in evolution or creationism; instead, they wanted to use the Scopes Trial and the teaching of evolution "as a parable, a metaphor for any kind of mind control," adding that "it's not about science versus religion. It's about the right to think" (Blankenship, 2001). The authors included a disclaimer noting that their play "does not to pretend to be journalism. It is theatre. It is not 1925.... It might have been yesterday. It could be tomorrow."

Lawrence and Lee shopped their play to eight Broadway producers. After being rejected by them all, Lawrence and Lee premiered *Inherit the Wind* on January 10, 1955, at Margo Jones's Theatre '55 in Dallas, Texas. The play garnered great reviews, prompting Lee to declare, "Margo [Jones] saw in *Inherit the Wind* the conflict and the drama which not one other person in American theatre saw" (Cattarulla, 2005). On April 21, 1955, *Inherit the Wind*—which was promoted as being "based on the thrillingly penetrating Scopes Trial"—opened at Broadway's National Theatre. When its run ended on June 22, 1957, after 806 performances, it was Broadway's most successful and longest-running drama. John Scopes turned down his invitation to attend the Broadway play, claiming that he too busy and not interested enough to fly to New York to see it. *Inherit the* Wind, which has been translated into more than 30 languages, became one of the most-performed plays in the world.

In 1956, director Stanley Kramer bought the screen rights to *Inherit the Wind*. Although the play *Inherit the Wind* made money, Kramer's film adaptation of *Inherit the*

Wind in 1960 lost money. Nevertheless, Kramer's movie became the source of much of the public's "knowledge" about the Scopes Trial.

In 1965, Lawrence and Lee founded the American Playwrights Theatre to bypass Broadway as a staging site for new plays. In addition to his collaborations with Lee, Lawrence also wrote independent manuscripts, including his acclaimed *Actor: The Life and Times of Paul Muni* (1974). Lawrence's work with Lee continued until Lee's death.

In 1986, The Ohio State University opened the Jerome Lawrence and Robert E. Lee Theatre Research Institute in Lawrence and Lee's memory. Lawrence and Lee won a variety of awards, including the *Variety* Critics Poll New York (1955), *Variety* Critics Poll London (1960), and the Donaldson Award in 1955 for *Inherit the Wind*. Lawrence and Lee were awarded the Lifetime Achievement Award by the American Theatre Association in 1979, and were elected to the Theatre Hall of Fame in 1990.

Later in his life, Lawrence taught playwriting at the University of Southern California. He died at age 88 on February 29, 2004, in Malibu, California, after a stroke. Lee died of cancer on July 8, 1994, in Los Angeles, California, where he is buried in Forest Lawn-Hollywood Hills Cemetery. Lee's tombstone includes a line from *Inherit the Wind*: "An idea is a greater monument than a cathedral."

Sources

Blankenship, Bill. 2001. Inherit the controversy. *Topeka Capital-Journal*, March 2, 19.

Cattarulla, Kay. 2005. The Texas Tornado. *Legacies, A History Journal for Dallas and North Central Texas*, 16 (2), 39–47.

Eisler, Garrett B. 2007. Lawrence, Jerome (1915–2004), and Robert E. Lee (1918–1994). In Cody, Gabrielle H., and Evert Sprinchorn (Eds.). *The Columbia Encyclopedia of Modern Drama*, Vol. 1. New York: Columbia University Press, 801–802.

Laws, Curtis Lee
1868–1946

Curtis Lee Laws was born in 1868 in Loudon County, Virginia. After graduating from Crozer Theological Seminary in 1893, he pastored Baptist churches in Baltimore (1893–1908) and Brooklyn (1908–1913). He then became editor of the *Watchman-Examiner*, which he built into the largest weekly Baptist newspaper in the North (Bradbury, 1946).

By the late 1910s, Laws was concerned about the growing popularity of modernism in Baptist churches and seminaries. On June 20–22, 1920, just before the annual meeting of the Northern Baptist Convention in Buffalo, New York, Laws convened a "General Conference on the Fundamentals" to promote theological orthodoxy and "restate, reaffirm, and reemphasize the fundamentals" of the faith. Weeks later, Laws coined the term *fundamentalist* when he wrote that people "who cling to the great fundamentals and who mean to do battle royal for the fundamentals shall be called 'Fundamentalists" (Laws, 1920). To Laws, fundamentalism was a protest against modernist interpretations of Christianity that discredited the supernatural. Laws' broad definition, which did not require biblical inerrancy, caught on; within a few years, the term *fundamentalist* united conservative Baptists, Lutherans, Catholics, Pentecostals, Ku Klux Klansmen, and others.

Laws did not attend the Scopes Trial, but a month after it ended, he complained that the trial "ought never to have been made an issue of fundamentalism." Laws then warned that "fundamentalists will be wise to major on another matter than evolution" (Laws, 1925), but the fundamentalists weren't listening.

Laws resigned as editor of the *Watchman-Examiner* in 1938, after which he helped organize Eastern Baptist Theological Seminary (now Palmer Theological Seminary) in Philadelphia and the Association of Baptists for World Evangelism. Laws died in his home in Manhattan, New York, on July 7, 1946. He is honored by the Curtis Lee Laws Memorial Chapel at Palmer Theological Seminary.

Sources

Bradbury, J.W. 1946. Curtis Lee Laws, D.D., L.L.D.: An Appreciation. *Watchman-Examiner*, 34, 29.
Laws, Curtis Lee. 1920. Convention sidelights. *Watchman-Examiner*, July 1, 834.
———. 1925. Editorial notes and comments. *Watchman-Examiner*, August 20, 1071.

Malone, Dudley Field
1882–1950

David Dudley Field Malone was born in New York City's Hell's Kitchen on June 3, 1882. He was named for prominent attorney Dudley Field (1805–1894), for whom Malone's father worked. Malone graduated from Chelsea's College of St. Francis Xavier in 1903 and, two years later, from Fordham Law School. He began practicing law in 1907, worked as an assistant counsel of New York City, helped found the NAACP, and became active in New York's Democratic Party. About this time, he dropped his "David" name because it would not fit on his office door's nameplate (Reddy, 1951). In Dayton, Malone was "astonished that the Dayton people can be so decent, sympathetic, and kind with those who support a proposition to which they are so much opposed" (Bailey, 1953).

On April 21, 1912, after helping Woodrow Wilson's successful campaign for president, Malone was appointed Third Assistant Secretary of State under Secretary of State William Jennings Bryan. Seven months later, Wilson appointed Malone the Collector of the Port of New York, a coveted patronage position having a salary of $12,000 per year. Malone became an advocate for women's suffrage and resigned as Collector on September 7, 1917, to protest Wilson's slow support of women and a Woman Suffrage Amendment to the U.S. Constitution. As Malone told Wilson, "I cannot and will not remain in office and see women thrown in jail because they demand their political freedom" (Dudley Field Malone, 1, 2013). The first public statement by Malone's replacement, writer Byron Newton (1861–1938), condemned women's suffrage.

In 1920, Malone ran for governor of New York on the newly formed Farmer-Labor Party ticket, but received only 69,908 of 2.8 million (i.e., less than 2.5 percent) votes. Thereafter, Malone focused on his international law business, which specialized in fast divorces for rich clients (e.g., Gloria Swanson). The colorful Malone, who was studied as a model of success (Fowler, 1906), was a popular speaker and social arbiter who told "the president and members of the Cabinet just how to act" (This social arbiter, 1913). A few months after his first marriage to May Patricia O'Gorman (1884–1961) ended in 1921, Malone married writer and suffragette Doris Stevens (1888–1963), a leader in the National Woman's Party. Malone, one of the leading male supporters of women's

Dudley Field Malone (left) and Clarence Darrow (right) were John Scopes' (center) most prominent defenders in Dayton (RCHGS).

suffrage, often defended women arrested for demanding the right to vote. In Dayton, Stevens created controversy when she registered at Hotel Aqua—which John Scopes described as "the only real hotel in town"—under her maiden name instead of as Doris Malone or Mrs. Dudley Malone. Stevens noted, "I came all the way from New York to find that the defendant was a man, the prosecutor was a man, the judge was a man, the jury all men, the attorneys on both sides men, the witnesses for the defense carefully selected men of intelligence. One would think there weren't any women in this world, or that they didn't do any thinking" (Wife of Malone, 1925).

The dapper, "noticeably silk-shirt-conscious" Dudley Malone, the only professing Christian on Scopes' defense team, impressed Dayton's residents with his wit and exquisite clothes, but was sometimes out-dressed by Jack Thompson, the jury foreman (Darrow, 1932). The patronizing, paternal Malone was "rather unpopular" among most members of the defense team and expert witnesses (Potter, 1951).

Before and during the trial, Malone—who denounced fundamentalism as the "blind faith of bigots"—promoted Henry Fairfield Osborn's claim that "man did not ascend or descend from the ape, but from a lower form of life of his own" and that "both theories should be taught" (Says legislature, 1925). Malone was relatively quiet in court until the fifth day (i.e., Thursday, July 16). After Bryan gave a rousing speech, Darrow whispered to Malone that "I'm an agnostic and Hays is a Jew. You've got to answer it"

(Reddy, 1951). Malone then delivered his "There is never a duel with the truth" speech defending the inclusion of expert testimony by scientists that the defense had brought to support Scopes and the teaching of evolution:

> Are we to have our children know nothing about science except what your church says they shall know? I have yet to see harm come to the young from learning and understanding, from intellectual humility and open-mindedness.... Mr. Bryan is not the only one who has a right to speak for the Bible.... Keep your Bible in the world of theology where it belongs but do not try to tell an intelligent world ... that the books of the Bible, written by men who knew none of the accepted facts of science, can be imposed upon a course of science.... There is never a duel with the truth. The truth always wins—and we are not afraid of it. The truth is no coward. The truth does not need the law. The truth does not need the forces of Government. The truth does not need Mr. Bryan... [Malone, 1929].

According to H.L. Mencken, Malone's words "roared out of the courtroom's open windows like the sound of artillery," and the court stenographer noted that the speech was followed by "profound and continued applause." Scopes considered Malone's speech to be the turning point of the trial and, more than 30 years later, described it as "the most dramatic event I have attended in my life ... [something] so dramatic that a transcript couldn't tell it." (Charles River Editors, 2016). John Butler, who had drafted the legislation used to prosecute Scopes, praised Malone's speech as "the finest speech of the century," and Bryan—a famous orator himself who had given and heard hundreds of impressive speeches—lauded Malone's speech simply as "the greatest speech I've ever heard." The *New York Times* claimed that the Scopes Trial had produced "the greatest debate on science and religion in recent years," and the *Nashville Banner* announced that "all prophesies that the trial of the Rhea county school teacher would rank with that of Galileo are well "night fulfilled" (Perry, 1925). Malone later claimed that his famous speech in Dayton was "the only truly extemporaneous speech [I] ever made" (Malone, 1929).

When Bryan died in Dayton five days after the trial, Malone—unlike Mencken and many others—praised his former boss as someone who had done more than anyone "to establish certain standards of conduct in politics." Two months later, however, Malone's response to Bryan's undelivered "Last Message" denounced Bryan as "the ablest leader, from the most sinister movement in the United States" (Malone, 1929). Malone also condemned the Butler Act as "a daring attempt to determine the truth or falsity of scientific thinking by the act of a legislature" and noted that "the issue at Dayton was ... the Constitutional guaranty of religious freedom and complete separation of Church and State.... Infidelity will not come from the study of science or of constantly revealed facts. Infidelity will come to haunt the aisles of Faith only when children grown to maturity realize that they have been denied access to full knowledge by any church or dogma." Malone then quoted Princeton biologist Edwin Conklin: "If evolution is false, it cannot be saved by science; if it is true, it cannot be destroyed by theology" (Malone, 1929).

After the trial, Malone remained a popular after-dinner speaker and a highly sought advisor on social etiquette. He considered his speeches "my sole medium of creative expression," and, in *"Unaccustomed as I Am"—Miscellaneous Speeches* (1929), he admitted to being "so vain as to print them." On April 20, 1931, Malone and Hays starred in a radio broadcast titled "The Trial of Napoleon Bonaparte," in which Malone defended Bonaparte and Hays led the prosecution. (Bonaparte was exonerated by a group of judges that included Bryan's daughter, Ruth Bryan Owen). Malone—whose friends included J.P. Morgan, H.G. Wells, and William Randolph Hearst—also endorsed commercial products such as Pabst "Blue Ribbon Genuine Ale," whose national advertisements described him as "a brilliant personality ... orator, world traveler, famous lawyer, cosmopolitan, New Yorker *par excellence* ... a keen observer of life and an uncompromising seeker of distinguished quality.... Dudley Field Malone says it's good, *that's news*!" (A brilliant personality, 1934). When later talking about Malone, Arthur Garfield Hays—Malone's law partner for many years—often noted that "I have never known a man of more personal charm" (Reddy, 1951).

Malone enjoyed his success, wealth, fame, and celebrity. He financed Gertrude "Trudy" Ederle's 1926 swim of the English Channel, represented New York at the Democratic National Convention in 1932, arranged exhibition matches for heavyweight boxing champion Gene Tunney, and joined national controversies such as the Lindbergh kidnapping in 1932. But behind the scenes, there were problems. After earlier advocating women's issues, Malone tired of Stevens and the women's movement. He refused to share his wealth with Stevens or pay money that he owed her, all while using her feminist principles to justify his many sexual liaisons with other women. Malone humiliated, cursed, threatened, and violently beat Stevens, something that Stevens later described as "a devastating experience" (White, 1993). Three months after Malone and Stevens divorced in Paris in October 1929, Malone married actress Edna Louise Johnson. The next year, they produced Malone's only child, Dudley Field "Shim" Malone, Jr. (1931–1990), who became a theatrical agent and manager.[23] The Malone-Johnson marriage lasted almost 10 years.

By the mid–1930s, divorces had become easier and cheaper to obtain, and Malone's divorce business declined sharply. In 1935, after stating, "I have had a tragic year," Malone filed for bankruptcy, citing assets of $62,500 and liabilities exceeding $260,000 (Dudley F. Malone, 1935). Malone then moved to Los Angeles, California, where he occasionally worked as a bit actor and lawyer for Twentieth Century–Fox studios. Because of his resemblance to his friend Winston Churchill, Malone's most prominent acting role was as Churchill in Warner Bros. Entertainment's *Mission to Moscow* (1943). Malone also supplied Churchill's voice in *Edge of Darkness* (1943) and played Churchill again (without credit) in *An American in Paris* (1951), which starred Gene Kelly (who in 1960 would appear in *Inherit the Wind*). Much of the time, however, Malone was unemployed.

In July 1949, while living in Los Angeles, a roadside attack by three brothers left Malone gashed and bloody. Just a few months later, Malone—who was then living with radio producer John Masterson—was "virtually penniless." Admitting that "there are new battles now, but a new generation will have to fight them," Malone drifted into obscurity (Reddy, 1951). Malone and Mencken had considered "doing a joint record" of the Scopes Trial, but they never did (Bode, 1977).

On October 4, 1950, the 68-year-old Malone entered Culver City Hospital, where he died the next day of a heart attack. Although his death was reported on the front pages of numerous newspapers, Malone was buried in an unmarked grave in Holy Cross Cemetery in Culver City, California. Later proceedings confirmed that Malone had assets totaling only $114.34, which was his unpaid salary from Twentieth Century–Fox (Dudley Field Malone believed broke, 1950). Malone's obituary remembered him as someone "known on two continents for his charm as a raconteur and for his wit as a speaker and phrasemaker" who was "once noted as an international lawyer" (Dudley F. Malone dies, 1950). Malone's lengthy, handwritten autobiography was lost.

Sources

Bailey, Kenneth Kyle. 1953. The Antievolution Crusade of the Nineteen-twenties. Ph.D. Thesis, Vanderbilt University, Nashville, Tennessee, 153.
Bode, Carl. 1977. *The New Mencken Letters*. New York: Dial Press, 241.
Brilliant personality applauds a real ale. 1934. *Fortune*, 10 (6) (December), 210.
Charles River Editors. 2016. *The Scopes Monkey Trial*. Newark, NJ: Charles River Editors, 25.
Dudley F. Malone dies in California. 1950. *New York Times*, October 6, 1.
Dudley F. Malone files as bankrupt. 1935. *New York Times*, December 22, 13, accessed September 6, 2020, at https://www.nytimes.com/1935/12/22/archives/dudley-f-malone-files-as-bankrupt-former-collector-of-port-fixes.html.
Dudley Field Malone, 1: The courage of his convictions. 2013. *Postscripts*, February 18. Accessed September 1, 2020, at http://notorc.blogspot.com/2013/02/dudley-field-malone-1-courage-of-his.html
Dudley Field Malone believed broke at death. 1950. *Los Angeles Times*, December 1, 40.
Fowler, J.A. 1906. Mr. Dudley Field Malone: A practical demonstration or lesson in phrenology. *The Phrenological Journal*, 119 (1), 78–79.
Malone, Dudley Field. 1929. *"Unaccustomed as I am"—Miscellaneous Speeches*. New York: J.J. Little and Ives Company.
Perry, Ralph. 1925. Eventful hour of Scopes Trial. *Nashville Banner*, July 17, 1.
Potter, Charles Francis. 1951. *The Preacher and I: An Autobiography*. New York: Crown.
Reddy, John. 1951. Magnificent failure. *Esquire*, 36 (2), 29.
Says legislature can't make morals. 1925. *Nashville Banner*, June 28, 5.
Scopes, John T., and James Presley. 1967. *Center of the Storm: Memoirs of John T. Scopes*. New York: Holt, Rinehart and Winston, 156, 155.
This social arbiter informs Washington officials on actions. 1913. *The Joplin (MO) Daily Globe*, August 21, 2.
White, Kevin. 1993. *The First Sexual Revolution: The Emergence of Male Heterosexuality in Modern America*. New York: New York University.
Wife of Malone puzzled at trial. 1925. *Baltimore Evening Sun*, July 17, 2.

The Mansion
1886–1945

The Mansion at the southwestern edge of Dayton was a large, elegant, Victorian house completed on December 10, 1884, by Dayton Coal and Iron Company (DCIC) for the company's owners and superintendent. Its 25 rooms included 10 bedrooms, expensive Victorian furniture, and 12 imported marble fireplaces. But after the collapse and bankruptcy of DCIC in 1913, the house was abandoned and soon was a "gray, bleak shell of its former grandeur." The empty house became a popular site of teenage parties and dances, some of which were attended by John Scopes. Some people believed the house was haunted by ghosts of DCIC miners killed in explosions.

In 1925, George Rappleyea and his wife "Precious"—suspecting that local residents would not want to house the "heretics" coming to town to help Darrow—renovated the Mansion for the defense team's witnesses. Despite these improvements, the facilities remained crude; camp beds, used mattresses, and second-hand furniture were adequate, but the house's electricity and water supply often failed. Despite these problems, the Mansion was a popular, boisterous gathering place during the trial.

In 1927, Walter E. Brewer—the father of trial participant Tommy Brewer and the general manager of Cumberland Coal & Iron Company—moved his family into the Mansion, where they lived until 1944. The Mansion burned down in 1945 (Ellis, 1945).

The Mansion was built in 1884 atop a hill overlooking Dayton to house the general manager and guests of Dayton Coal & Iron Company. In 1925, George Rappleyea and his wife, Ova "Precious" Rappleyea, refurbished the abandoned Mansion for attorneys and guests of John Scopes' defense team (RCHGS).

Source

Ellis, Jacob. 2015. Coconut Grove: The role of a haunted mansion in the Scopes Trial. *Rhea Notes*, 22 (4), 1–3.

Martin, Thomas Theodore "T.T." 1862–1939

Thomas Theodore "T.T." Martin was born in Smith County, Mississippi, on April 26, 1862. After graduating from Mississippi College in 1886, he taught science (e.g., at what is now the University of Mary Hardin–Baylor) before being ordained a Baptist minister in 1888. Martin earned a degree from the Southern Baptist Theological Seminary and later became the most active Southern Baptist in the fundamentalists' campaign to ban the teaching of evolution in public schools.

When William Bell Riley's Anti-Evolution League of Minnesota became the Anti-Evolution League of America in 1924, Martin was named the organization's field secretary and editor of its newspaper, *The Conflict*. Martin told crowds that after being taught about evolution, a "child's faith in the Savior is gone forever, and her soul is doomed for Hell, and with your taxes, you paid to have it done," and that evolution was "sweeping our sons and daughters away from God ... to spend eternity in hell," and that "if [evolution] is true, then Jesus Christ was a bastard, illegitimate son of a fallen woman, not Deity, not really God's son, not really our Redeemer and Savior at all..." (Martin, 1923a, b). Then, invoking Bryan's famed "Cross of Gold" speech from 1896,

Evangelist T.T. Martin was the loudest fundamentalist to come to Dayton for the Scopes Trial. Here, he posts a sign on a bathroom telling people to "Read Your Bible" (RCHGS).

Martin warned that evolution's "pagan followers shall not press their crown of paganism on the brow of American youth; they shall not crucify the souls of our children on the cross of their veneered infidelity" (Thompson, 1975).

In 1923, Martin published his popular, 174-page *Hell and the High Schools* (1923), which urged parents to save their children from "spending eternity in hell" by pressuring school boards and legislators to ban the teaching of evolution. Throughout his book, Martin claimed that although "many do die from small-pox and yellow fever, many, many [more] … are being damned eternally by the teaching of Evolution in our schools.… Ramming poison down the throats of our children is nothing compared with damning their souls with the teaching of evolution.… [Evolutionists] laugh and jeer, as the rapist laughs and jeers at the bitter tears of the crushed father and mother over the blighted life of their child." Like other fundamentalist leaders, Martin linked the teaching of evolution with German militarism, noting, "The Germans who … fed little children poisoned candy were angels compared to the teachers, paid by our taxes, who fed our children's minds with the deadly, soul-destroying poison of Evolution.… Evolution and the teaching of Evolution in tax-supported schools is the greatest curse that ever fell upon this earth" (Martin, 1923a). In Dayton, Martin—the only antievolutionist of any note aside from Bryan to attend the Scopes Trial—rented a store in which he sold his and other fundamentalists' books.

On the afternoon of Sunday, July 12, in Dayton, Martin was scheduled to speak on the dais on the courthouse lawn on what was rumored to be "Shall we be taxed to damn our children?" However, when Unitarian preacher Charles Potter's sermon at nearby First United Methodist Church (North) was cancelled, Martin yielded his time to Potter, provided Potter did not peach about evolution. Potter preached about brotherly love and the importance of a liberal education for establishing one's values.

After the Scopes Trial ended, Martin stayed in Dayton and conferred with Bryan about how "to save our schools from the curse of evolution" (Martin, n.d.). In February 1926, Martin lobbied the Mississippi House of Representatives to pass a bill similar to Tennessee's Butler Act, daring legislators opposing the bill to "go back to the fathers and mothers of Mississippi and tell them ... [that] you turned their children over to a teaching that God's Word is a tissue of lies" (Thompson, 1975). The bill eventually passed and was signed into law by Governor Henry L. Whitfield. Mississippi, Tennessee, and Arkansas were the only three states to criminalize the teaching of evolution in their public schools.[24] To the end of his life, Martin claimed, "The resurrection of the Savior kills evolution" and that "it is, in the sight of God, a crime to teach [evolution] in any school" (Martin, n.d.).

Martin died on May 23, 1939, in Baptist Hospital in Jackson, Mississippi, and was buried in Roseland Cemetery in Gloster, Mississippi. Martin's tombstone is inscribed with John 3:16, Titus 2:13–14, and John 5:24, the Bible verses on which he based his life.

Sources

Martin, T.T. 1923. *Hell and the High Schools: Christ or Evolution Which?* Kansas City: Western Baptist Publishers Co.

______. 1923b. *The Evolution Issue*. Los Angeles: n.p., 34.

______. n.d. Viewing Life's Summit from Pikes Peak: Life Story of T.T. Martin. Louisville, KY: A.D. Muse, 31, 78, 81.

Thompson, James J., Jr. 1975. Southern Baptists and the anti-evolution controversy of the 1920s. *Mississippi Quarterly*, 29 (1), 65–81.

McKenzie, Benjamin Gordon 1866–1938

Benjamin Gordon "Ben" McKenzie was born on February 26, 1866, in Meigs County, Tennessee. After attending schooland becoming interested in debate, McKenzie returned to Meigs County and taught school for $15 per month. He was admitted to the bar in 1887 in Decatur, where he practiced law for six months, and then moved to Dayton.

In 1898, McKenzie was elected to represent Rhea County in Tennessee's 51st General Assembly. During his one term in the legislature, McKenzie introduced House Bill 210, which required companies issuing scrip as pay to redeem the scrip with money upon demand. On March 7, 1901, McKenzie—in a rented tuxedo—argued this case before the U.S. Supreme Court (*Dayton Coal & Iron Company v. Barton*, 183 U.S. 23, 1901). McKenzie was so impressed with Chief Justice Oliver Wendell Holmes that he returned to Dayton and changed the name of his youngest son (George Jeremiah, born the year before) to Oliver Wendell McKenzie (McKenzie, 2006).

In January 1915, Tennessee created the Eighteenth Judicial District, for which

Tennessee Governor Thomas C. Rye appointed McKenzie the district's first Attorney General. McKenzie was elected without opposition until 1924. McKenzie, "The Dean of the Rhea County Bar," then practiced law in Dayton with his son, Gordon. McKenzie was a well-respected attorney and popular after-dinner speaker.

At the Scopes Trial, McKenzie and Bryan were charged with preparing the summation, but McKenzie didn't wait until then to speak. At one point, McKenzie claimed that it was "better to kill all children under two years of age than to have teachers who will wreck the lives of children with atheist theories which will put them on the toboggan slide which leads to hell." At a rally during the trial, McKenzie invoked regional pride, denouncing "the Northern lawyers, who had come down to teach the 'ignorant yokels' what to believe, they had better go back to their homes, the seat of thugs, thieves, and Haymarket rioters,[25] and educate their criminals rather than to try and proselyte here in the South, where people believe in the Christian religion, and know that Genesis tells the full and complete story of creation" (Linder, 2020). On July 17, while arguing against the admissibility of scientific evidence, McKenzie described Tennessee's view of evolution as "God throwing a dishrag into the ocean. Life evolves from this. The Master then said, 'Old Boy, in 6,000 years I will make you something'" (State's view of evolution, 1925). Later, McKenzie described the trial as "a fake, fraud and farce that never convinced anyone that we were men or monkeys" (Broyles, 1991).

After the trial, Darrow offered McKenzie's son Gordon a job with his law firm in Chicago, but the elder McKenzie wanted Gordon in Dayton. McKenzie, who exchanged letters with Darrow until their deaths, described Darrow as "a champion of the underdog and a strong Democrat. A charitable man and one who is attuned to life, with which he is completely reconciled." McKenzie dismissed Mencken's insults as "just shooting hot air for cash," but was impressed when Mencken admitted that he was surprised "that one of my [i.e., McKenzie's] intelligence would disbelieve in evolution" (McKenzie, 2006).

In February 1926, McKenzie was arrested for "transporting an intoxicant" (i.e., whiskey). Darrow offered to defend him, but Ben declined, claiming that the arrest was a political ploy meant to discredit his son Gordon, who was running for re-election for county judge against a resident of Spring City. McKenzie's $50 fine was dismissed. When Darrow published *The Story of My Life* in 1932, he sent McKenzie one of the first copies, inscribed "to my eternal and valued friend, B.G. McKenzie—with the kindest remembrances" (McKenzie, 2006).

When Darrow returned to Dayton in 1930, he met Ben at Robinson's Drug Store for a picture at the table where the trial originated. In August 1933, Ben took his family to the World's Fair in Chicago, where he visited with Darrow at his home. This was the last time they saw each other. Five years later, both died within three months of the other.

McKenzie's last public appearance was in May 1938 in Chattanooga, where he attended a ceremony honoring Congressman Sam McReynolds. On June 23, 1938, while at his office in downtown Dayton, McKenzie suffered a stroke. A second stroke followed, and a final one killed him on June 27, 1938. His pallbearers included F.E. Robinson, an instigator of the Scopes Trial; Scopes Trial Judge John T. Raulston was an honorary pallbearer. McKenzie is buried in a family plot in Dayton's Buttram Cemetery beneath the epitaph, "His greatest joy in life was in serving his fellow-men." The McKenzie Law Firm, which Ben founded in 1888, closed in 1998.

Sources

Broyles, Bettys J. 1991. *History of Rhea County, Tennessee*. Dayton, TN: Rhea County Historical and Genealogical Society, 315.
Linder, Douglas O. 2020. Scopes "Monkey" Trial: Ben G. McKenzie. Accessed September 19, 2020, at https://-famous-trials.com/scopesmonkey/2093-mckenzie.
McKenzie, James W. 2006. The role of law in society: How the Scopes Trial affected my family. Unpublished manuscript dated March 2006 from James McKenzie.
State's view of evolution. 1925. *New York Times*, July 17, 1.

McKenzie, James Gordon 1893–1944

James Gordon McKenzie was born on September 24, 1893, in Dayton, Tennessee, to Benjamin Gordon and Louella Foust McKenzie. In June 1916, after graduating from Cumberland University Law School in Lebanon, Tennessee, McKenzie began practicing law in Dayton with his father. On October 23, 1923, Tennessee governor Austin Peay appointed McKenzie to complete the term of Rhea County Judge David Rhea, a position in which he served until 1934.

In July 1925, McKenzie helped his father and Attorney General Tom Stewart prosecute John Scopes. Gordon, the custodian of the Rhea County Courthouse courtroom, approved all of the press-passes for reporters covering the trial. Before the grand jury met to indict Scopes, Gordon told reporters, "The teaching of evolution in public schools is detrimental to public morals, storms the very citadel of our Christian religion, repudiates God, and should not be permitted" (Scopes evolution law, 1925). On July 16, McKenzie arranged for some electric fans to ventilate the courtroom. Four days later, after announcing, "I believe in the Bible as strong as anybody else here," Gordon supported the suggestion to take down the "Read Your Bible" sign, to which his father Ben responded, "I do not agree with my son." The younger McKenzie later withdrew his statement, suggesting that the defense might "represent a force that is aligned with the devil and his satellites."

In 1934, Gordon was appointed a special attorney for the U.S. Department of Justice, during which he acquired land for Great Smoky Mountains National Park and Oak Ridge National Laboratory. McKenzie remained a prominent attorney and Democratic leader in Rhea County (Gordon McKenzie, 1944).

McKenzie died at his home of a heart attack on the morning of September 24, 1944, his 51st birthday. He is buried in the McKenzie family plot in Dayton's Buttram Cemetery.

Sources

Gordon McKenzie is dead in Dayton. 1944. *Chattanooga Times*, September 24, 3.
Scopes evolution law wholesome statute. 1925. *Chattanooga Times*, May 24, 9.

McPherson, Aimee Semple 1890–1944

Aimee Semple McPherson was born on October 9, 1890, in Salford, Ontario, Canada. As a child she preached to her dolls, and as a teenager wrote a letter to a Canadian newspaper questioning the use of tax money to support the teaching of evolution in public schools (Sutton, 2007). "Sister Aimee" continued to attack the teaching of evolution for the rest of her life. McPherson, like William Jennings Bryan, claimed that evolution produced a "survival of the fittest" society that made charity and social reform pointless.

In 1908, McPherson married Robert James Semple, but he died soon thereafter during a missionary trip to China. McPherson and her second husband, who lived in their "Gospel Car," divorced in 1921, and McPherson divorced her third husband in 1934. Throughout these setbacks, McPherson continued to preach, often in tent revivals that attracted growing crowds. In 1917, she founded *Bridal Call*, a magazine that discussed women's roles in religion. After moving to Los Angeles in 1918, McPherson built her Angelus Temple into an early megachurch that seated 5,300 people and boasted the largest single Christian congregation in the world. At its peak, McPherson's Angelus Temple enrolled almost 10 percent of the population of Los Angeles as members (Thomas, 1970).

McPherson, who preached up to 22 sermons per week, was the first woman to preach wirelessly on radio and the first woman to be awarded an FCC license (the call letters of her radio station, KFSG, stood for Foursquare Gospel, the religious denomination that she founded). Between 1919 and 1922, her revivals broke Billy Sunday's records by reaching more people than any touring theater or political production in American history (Epstein, 1993). Although some reporters portrayed McPherson as a sexual vixen, all acknowledged her immense appeal; as one reporter noted, McPherson, "has magnetism such as few women since Cleopatra" (Ryan, 1923). The flower-loving McPherson also sponsored and rode atop award-winning floats in Pasadena's Tournament of Roses Parade, from which she invited people to attend her nearby church.

McPherson led the fundamentalists' antievolution crusade in the West. She regularly hosted fundamentalist leaders such as Bryan and lamented how evolution was "poisoning the minds of the children of the nation." McPherson urged her followers to boycott schools that were teaching evolution and "sacrifice science rather than religion." She blamed Darwin for America's moral decline, denouncing evolution as "the greatest triumph of Satanic intelligence in 5,931 years of devilish warfare" that was responsible for "jazz, bootleg booze, the crime wave, student suicides … and the peculiar behavior of the younger generation" (Sutton, 2007).

Movie stars frequented McPherson's church, and some—including Charlie Chaplin—helped her design church services that rivaled those in nearby Hollywood. McPherson's unapologetically patriotic services, which anticipated those of television evangelists that were to come, often depicted villains such as Mussolini, Darwin, and Hitler, after which McPherson would emerge in an all-white gown and read the national anthem. For fundamentalists like McPherson who believed that humans were the apex of creation, and that Americans were the apex of humanity, Darwinism tarnished what it meant to be American.

Just before the Scopes Trial began, McPherson telegraphed Bryan that "ten thousand members of Angelus Temple with her millions of radio church members send grateful appreciation of your lionhearted championship of the Bible against evolution and throw our hats in the ring with you." She promised an all-night prayer service, a Bible-based parade through Los Angeles, and a church service that would culminate with the "hanging and burial of monkey teachers" (Sutton, 2003; McPherson to Bryan,

1925). After the trial, McPherson continued her attacks on evolution. However, on May 18, 1926, she disappeared. After reappearing five weeks later in Mexico, McPherson claimed that she had been kidnapped, but many suspected she'd had a tryst with Kenneth Ormiston, a former employee. The story made news worldwide, and the *New York Times* printed about the same number of stories (just less than 100) about McPherson's troubles as it had the previous year about the Scopes Trial and Bryan's death (Sutton, 2007). All charges against McPherson were later dropped, and McPherson claimed that Satan had given direct orders for her kidnapping, which had resulted from her opposition to dancing, liquor, and evolution (Shipley, 1927). Although her reputation suffered, McPherson remained popular; in her last national revival tour in 1933–1934, her 336 sermons were heard by two million people. Unlike many fundamentalist preachers, McPherson embraced integrated services at her church.

McPherson, the most publicized Protestant evangelist of her time, died alone in a hotel room in Oakland, California, on September 27, 1944, of a drug overdose. More than 45,000 people visited her casket in Angelus Temple to pay their respects. She is buried in a large, hillside plot in Forest Lawn Memorial Cemetery in Los Angeles, California. The church she founded became its own denomination, the International Church of the Foursquare Gospel, which in 2022 claimed more than eight million members. McPherson has been the subject of countless documentaries, books, musicals, plays, papers, and films.

Sources

Epstein, Daniel Mark. 1993. *Sister Aimee: The Life of Aimee Semple McPherson*. Orlando: Harcourt Brace, 111.

McPherson, Aimee Semple, to William Jennings Bryan, July 12, Box 47, Papers of William Jennings Bryan, Library of Congress, Washington, D.C.

Ryan, Don. 1923. Sister McPherson. *Los Angeles Record*, January 2.

Shipley, Maynard. 1927. *The War on Modern Science: A Short History of the Fundamentalists' Attacks on Evolution and Modernism*. New York: Alfred A. Knopf, 252.

Sister defied evolution. 1934. *Foursquare Crusader*, 8 (15) (January 24), 1.

Sutton, Matthew Avery. 2007. *Aimee Semple McPherson and the Resurrection of Christian America*. Cambridge: Harvard University Press.

Thomas, Lately. 1970. *Storming Heaven: The Lives and Turmoils of Minnie Kennedy and Aimee Semple McPherson*. New York: Morrow, 32.

Mencken, Henry Louis
1880–1956

Henry Louis "H.L." Mencken was born on September 12, 1880, in Baltimore, Maryland. After what he called a "placid, secure, uneventful and happy" childhood, Mencken graduated in June 1896 as the valedictorian of Baltimore Polytechnic Institute. He wanted to work for newspapers, but instead went to work in his family's tobacco business. Mencken took a free correspondence-class about writing in 1898 that was his only formal training in journalism, and after his father's death, he got a job with the *Baltimore Morning Herald*. Soon thereafter, he was working for the *Baltimore Sun*, to which he contributed articles until 1948. He also edited *The Smart Set*, which promoted writers such as Eugene O'Neill, F. Scott Fitzgerald, and Sinclair Lewis. In 1919, Mencken published *The American Language*, which described how English is spoken in the United States, and in 1924, he founded *The American Mercury*—"the magazine of cleverness."

Mencken's love of the written word was evident in his writing and throughout his career. He claimed that Mark Twain's *The Adventures of Huckleberry Finn* was the greatest piece of American literature. He also admired German philosopher Friedrich Nietzsche—something he shared with Darrow. In May 1925, when Mencken met with Darrow in Richmond to urge him to defend Scopes, expose fundamentalism, and "make a fool of Bryan," the two men were able to find some common ground. Mencken, however, was more outspoken in his questioning of representative democracy, which he believed enabled inferior people to oppress their superiors, and in his view that religion was a curse that damages thought and that its evidence for immortality was no better than that for witches.

In 1925, while with *The Baltimore Sun*, Mencken covered the Scopes Trial, which he famously dubbed "The Monkey Trial." More than 200 journalists covered the trial, but Mencken's syndicated columns had the most impact. Before he went to Dayton, Mencken often criticized fundamentalism, telling his readers that Genesis "offers, to an ignorant man, the irresistible reasonableness of the nonsensical. So he accepts it with loud hosannas, and has one more excuse for hating his betters." As the trial drew nearer, Mencken also defended Bryan's claim that citizens should control the curriculum of local schools, asking "how could it be otherwise? … No principle is at stake in Dayton save the principle that school teachers, like plumbers, should stick to the job that is set before them … the issue of free speech is quite irrelevant" (Mencken, 2006). Mencken also knew that "getting Scopes acquitted would be worth a day's headlines in the newspapers, and then no more, but smearing Bryan would be good for a long while" (Teachout, 2002).

While in Dayton, Mencken—who called himself the defense team's "Consulting Man of Vision"—wrote nine articles, all of which included his name in the headline to attract readers (Farrell, 2011). People in Dayton read Mencken's columns in the *Chattanooga News*. Mencken's first article (July 9) described Dayton as a "country town full of charm and even beauty," and his second article promised that Scopes "will be bumped off … to the applause of all right-thinking men." On Sunday, July 11, Mencken likened the trial to "a religious orgy," complete with "soul savers and sin exterminators," adding that the food served in Dayton's hotels is "tasteless and swims in grease." Subsequent articles described Darwin as "the devil with seven tails and nine horns," dismissed people in Dayton as "hillbillies," "peasants," "yokels," and "morons," and ridiculed Bryan as a "mangy and flea-bitten" person who "hates the learning he cannot grasp." After claiming that Judge Raulston believed "that in a trial for heresy it is perfectly fair and proper to begin the proceedings with a prayer for the confutation and salvation of the defendant," Mencken again denounced Bryan as a "mountebank" and "fundamentalist pope," while sarcastically adding that Scopes would get a "fair trial, as a fair trial is understood among fundamentalists." On July 17, Mencken praised Malone's rousing "We are not afraid" speech as "one of the best presentations of the case against fundamentalist rubbish that I have ever heard," and before leaving town the next day, told his readers that all that was left in Dayton "against the infidel Scopes is the formal business of bumping off the defendant" (Mencken, 2006). Mencken left Dayton after the first week of the trial, so he missed the famed confrontation between Bryan and Darrow. Mencken, who admitted that his reports from Dayton included "necessary unpleasantness," suspected that the trial would affect Americans for years to come, and he was right.

Although Bryan had once described Mencken as "the best newspaperman in the country," Mencken admitted that he despised Bryan, even adding, "I was against Bryan the moment I heard of him" (Mencken, 1993). When Bryan died five days after the trial, Mencken's hateful obituary—aptly called one of "the greatest masterpieces of invective in the English language" (Teachout, 2002)—described Bryan as a "dirty ... disgusting ... ignorant, bigoted, self-seeking, blatant, and dishonest ... pathetic fool [and a] walking malignancy ... [having a] vague, unpleasant manginess about his appearance.... All sense departed from him.... Like a dog with rabies ... a poor clod like those around him, deluded by a childish theology, full of an almost pathological hatred of all learning, all human dignity, all beauty, all fine and noble things. He was a peasant come home to the dung pile" (Mencken, 2006). The popularity of Mencken's columns is why John Scopes later admitted that his trial "was Mencken's show ... a mention of the Dayton trial more likely invokes Mencken than it does me." Scopes believed that Mencken's "biting commentary on the Bible Belt and the trial itself was one of the highlights of the entire event" (Scopes and Presley, 1967).

After the trial, Mencken remained a popular critic of pretense, Puritanism, and self-righteousness. In 1926 alone, he was the subject of more than 450 editorials (Rodgers, 1987). When Mencken was arrested that year in Boston for selling an issue of *American Mercury* that authorities charged was "sensational and pornographic," he was represented by Arthur Garfield Hays. Mencken married in 1930, and the following year—after dismissing Arkansas as the "apex of moronia"—the Arkansas legislature passed a motion to pray for Mencken's soul (Manchester, 1951). Mencken was also a strident racist, claiming that it is "a ridiculous waste of energy" to improve African Americans because "the educated Negro of today is a failure ... because he is a Negro." Mencken similarly dismissed Jews as "the most unpleasant race ever heard of" (Mencken, 1930). Mencken's columns evoked responses, and prominent authors such as Sinclair Lewis and Ernest Hemingway mentioned Mencken in their books; Lewis even dedicated *Elmer Gantry*—a novel about an unscrupulous, small-town preacher—to Mencken "with profound admiration."[26] Political commentator Walter Lippmann considered Mencken to be the "most powerful personal influence on this whole generation of educated people," and Scopes' defender Arthur Garfield Hays simply noted, "The press revered one god, Mencken" (Allen, 1931; Rodgers, 2005).

Mencken and Dudley Field Malone had talked about writing an account of the Scopes Trial, but they never did. Mencken later claimed that he'd been invited to attend the opening of William Jennings Bryan University in Dayton, but he did not go.

On November 23, 1928, Mencken suffered a stroke that left him unable to read or write, and barely able to speak. After Mencken died in his sleep in his home in Baltimore on January 29, 1956, the *Boston Globe* eulogized him as a "champion of the unfettered mind," to which Lippmann added that Mencken was a man "bigger than his ideas ... [and] a personal force in American life" who had "an extraordinarily cleansing and vitalizing effect." The *New York Times* claimed that the world was "impoverished by his death just as we were enriched by his life," while Mencken's *Baltimore Sun* lamented, "We shall not soon see his like again" (Teachout, 2002). Mencken's ashes were buried in a family plot in Baltimore's Loudon Park Cemetery. In 1983, Mencken's home—now a city museum—was made a National Historic Landmark. His *American Mercury* was later purchased by fundamentalists.

Sources

Allen, Frederick Lewis. 1931. *Only Yesterday: An Informal History of the 1920s.* New York: Harper & Row, 200.

Farrell, John A. 2011. *Clarence Darrow: Attorney for the Damned*. New York: Doubleday, 369.
Manchester, William. 1951. *Disturber of the Peace: The Life of H.L. Mencken*. New York: Harper, 252.
Mencken, H.L. 1930. *Treatise on the Gods*. New York: Knopf, 345–346.
______. 1993. *My Life as Author and Editor*. New York: Alfred A. Knopf, 34.
______. 2006. *A Religious Orgy in Tennessee*. Brooklyn: Melville House, 15, 21, 27, 35, 44, 50, 61, 63, 75, 82, 89, 103.
Rodgers, Marion Elizabeth. 1987. *Mencken and Sara: A Life in Letters*. New York: McGraw-Hill, 217.
______. 2005. *Mencken: The American Iconoclast*. Oxford: Oxford University Press, 275.
Scopes, John T., and James Presley. 1967. *Center of the Storm: Memoirs of John T. Scopes*. New York: Holt, Rinehart and Winston, 93.
Teachout, Terry. 2002. *Skeptic: A Life of H.L. Mencken*. New York: HarperCollins, 217, 221, 228, 332.

Mendi, Jo
c. 1920–1930

Jo Mendi was a chimpanzee bought in 1923 by circus owner Louis Backenstoe from an animal breeder in Cape Town, Africa. Backenstoe trained Jo to perform in circuses, fairs, and carnivals, as well as on vaudeville, Broadway, and in movies. Jo dressed himself and entertained crowds by dancing, skating, walking on tightropes, and smoking cigars. Jo's trainer and tutor was Gertrude Bauman (Backenstoe's wife), who allegedly insured Jo for $100,000 with Lloyd's of London (Backenstoe to collect, 1930). Jo, who was

Jo Mendi, a performing chimpanzee, was one of the most popular attractions in Dayton during the Scopes Trial. In this picture, Jo—under the watchful eye of his trainer and tutor Gertrude Bauman—gives a piano concert to some children on the front steps of the home of druggist F.E. Robinson (RCHGS).

advertised as having "the intelligence of a five-year-old child" (Joe Mendi, 2017), was the first famous chimpanzee trained to do human-like tricks.

In 1925, Bauman and Backenstoe brought Jo to Dayton for the Scopes Trial, telling reporters that everyone could use Jo to decide whether their "family tree bore cocoanuts or not." Jo and his humanlike behaviors were offered as evidence for the defense, but the offer was declined.[27] Mendi initially stayed at W.C. Bailey's boardinghouse (where John Scopes rented an upstairs room), but he was soon moved to the Ballard/Bailey house on Third Avenue. While in Dayton, Jo wore tailored suits as he posed for pictures, sipped Cokes at Robinson's Drug Store, played piano at the Robinsons' house, and shook hands with his fans. Many of the newspaper stories from the Scopes Trial mentioned Jo, such as this report that "in an attempt to reinforce the theory of evolution, a New York lawyer brought a trained chimpanzee to Dayton to show how closely related it was to humans. He took the chimpanzee to a diner for lunch. Dressed in a gray tweed suit and a red tie, the chimpanzee pulled out chairs and seated those at his table. The chimpanzee tucked a napkin in its shirt before dining and puffed a cigar after the meal, blowing smoke rings. 'This is a perfect example which proves man evolved from the ape,' the lawyer reportedly said.... This antic may have contributed to the fact that Dayton was later referred to as 'Monkey Town'" (Simpson, 2015). Several reporters noted that Jo was "much better behaved than most children" (Potter, 1951). After the trial, Jo's performance at Dayton was recalled as having caused "a veritable sensation and near-riot among the Darrow-Bryan factions at the historical Scopes evolution trial" (Created sensation at evolution trial, 1935).

After the Scopes Trial, Jo lived at the Detroit Zoo, where he performed as many as 14 shows per day. After getting sick weeks earlier in Peoria, Illinois, Jo died of ptomain poisoning in the Backenstoe home in Columbus, Ohio, on July 22, 1930. Jo's death was reported in newspapers across the country. His body was then preserved and displayed in a traveling exhibit (Backenstoe to collect, 1930).

Jo was soon replaced by Jo Mendi II, who in 1932 began performing at the Detroit Zoo. In all, there were four chimps named Jo Mendi, the last of which died in 2005. In the movie *Inherit the Wind* (1960), Jo was played by Jerry Hall, a two-and-a-half-year-old chimp.

Sources

Backenstoe to collect $100,000 on Joe Mendi. 1930. *Billboard*, August 2, 8.
Caused sensation at evolution trial. 1925. *The (Ottawa) Citizen*, September 3, 19.
Joe Mendi the gentleman chimpanzee made America swoon in the 1920s. 2017. HuffPost December 6, Accessed November 4, 2020, at https://www.huffpost.com/entry/joe-mendi-the-gentleman-chimp_n_4220823.
Potter, Charles Francis. 1951. *The Preacher and I: An Autobiography*. New York: Crown.
Simpson, Bob. 2015. This and that: Little known things about the trial. *Rhea Notes*, 22 (4), 7–8.

Metcalf, Maynard Mayo
1868–1940

Maynard Mayo Metcalf was born on March 12, 1868, in Elyria, Ohio. After earning an A.B. degree from Oberlin College in 1889 and a Ph.D. from Johns Hopkins University in 1893, he studied marine invertebrates at a variety of institutions, including Woods Hole Biological Laboratory and the National Museum of Natural History. Metcalf, like some of the other scientists brought to Dayton to help the defense, strongly advocated eugenics; he considered it "criminal" and "heinous" to produce children who "will be an injury to society," and denounced charities and social workers who helped "unfit" people reproduce as "accessory to the crime." Metcalf urged people to make eugenics their religion because he believed that eugenics could "banish physical, mental, and moral invalidism" (Metcalf, 1916).

Metcalf, a Congregationalist who taught Bible classes in Baltimore, was the only scientist to testify from the witness stand at the Scopes Trial. Metcalf testified on the afternoon of Wednesday, July 15, just after the state had rested its case. The defense hoped that Metcalf would show that an intelligent person could be both a Christian and an evolutionist, and therefore that evolution doesn't deny the Biblical story of creation. Metcalf testified "from personal knowledge that [scientists] feel and believe ... that evolution is a fact" and that disagreements among them about evolution relate only "to the exact method by which evolution had been brought about." Attorney General Tom Stewart objected to Metcalf's testimony, emphasizing that the nature of evolution was for the judge, not the jury, to decide. When Judge Raulston then excused the jury, he warned them not to "linger in the courthouse yard" where they might hear Metcalf's testimony on WGN's loudspeakers.

Metcalf concluded that it is "entirely impossible for any normal human being who was conversant with the phenomena to have even for a moment the least doubt" that humans evolved. His testimony, which Mencken praised as "one of the clearest, most succinct and withal most eloquent" explanations of evolution that he had heard, was the only time evolution was explained from the witness stand in Scopes' trial (Mencken, 1925). The next morning, Metcalf returned to the witness stand, but when Darrow asked his first question, Stewart again objected to Metcalf's testimony. The opposing attorneys spent the rest of the day arguing about the admissibility of the scientists' testimony. With the jury gone, Raulston admitted that the admissibility of the scientists' testimonies was "the most difficult thing the court is going to pass on."

On Friday morning, Raulston announced,"The court is content to sustain the motion of the Attorney General to exclude the expert testimony." According to Raulston, the Butler Act was clear to "the ordinary, non-expert mind" and its interpretation did not require the guidance of experts. When Hays objected and announced the defense team's intention to appeal Raulston's "denial of justice," Raulston agreed to let the defense submit written affidavits summarizing the testimony of the excluded experts. The defense team's scientists had no other roles in the trial.

A week after the trial, *Science* reported that the defense team's scientists were "impressed with Mr. Scopes's intellectual qualities and modesty, and believing he is entitled to some substantial recognition for the trying experience that he has undergone in the service of science, and of liberty of thought and speech ... have organized a committee to raise a scholarship fund of $5,000 [to] enable Mr. Scopes to continue his scientific training as soon as possible." Metcalf chaired The John T. Scopes Scholarship Fund which raised $2,550 that Scopes used to do graduate work at the University of Chicago. Despite claims in many books and papers, Scopes never earned a graduate degree.

From 1926 to 1933, Metcalf was a zoology professor and research associate at Johns Hopkins University. After retiring, he moved to Winter Park, Florida, where he died on April 19, 1940. Metcalf is interred there at Oaklawn Cemetery.

Sources

The John T. Scopes Scholarship Fund. 1925. *Science*, 62 (1596, July 31), 105.
Mencken, H.L. 1925. Mencken declares strictly fair trial is beyond ken of Tennessee fundamentalists. *The Baltimore Evening Sun*, July 16, 1.
Metcalf, Maynard M. 1916. Evolution and man. *Journal of Heredity*, 7 (8), 356–364.

Miller, Arthur McQuiston 1861–1929

Arthur McQuiston Miller was born on August 6, 1861, in Eaton, Ohio, where he spent much of his youth collecting fossils. He earned a B.A. degree (1884) and an M.A. degree (1887) from Princeton, and in 1892 became a professor of geology and zoology at what is now the University of Kentucky. At the urging of students, and despite his limited knowledge of the game, Miller became Kentucky State's inaugural football coach. After a 2–4–1 season, Miller relinquished his coaching job so he could focus on science.

By 1907, Miller headed Kentucky's geology department, and the next year he became the school's first Dean of Arts and Sciences. Miller, a fossil expert who had studied with geologist Henry Fairfield Osborn, taught John Scopes geology at Kentucky, and they remained friends for life. Miller, who inaccurately claimed that Bryan believed that Earth is only "a few thousand years" old, knew that geology is "absolutely contradictory to a six-day creation" because it documents "a continuous succession of life extending back millions of years." Miller likened attempts to teach biology and geology without evolution to presenting the play *Hamlet* "with Hamlet left out." Not content to restrict his comments to science, Miller also denounced Bryan as a "sorry spectacle of a man assuming the role of proficiency in a science" and "ignorant of [science's] fundamental facts and principles" (Miller, 1925).

In 1925, Dudley Field Malone announced that Miller would be in Dayton to testify for the defense (Several expert scientists, 1925). As the trial approached, Miller told reporters that "science cannot by limited by Biblical concepts" and that the Scopes Trial would be "a test to see if scientific facts and theories can be established by legislative enactment" (A.M. Miller may testify, 1925). However, like several other prominent scientists who ridiculed Bryan, Miller did not show up in Dayton to confront Bryan face-to-face.

On October 28, 1929, the 69-year-old Miller died while on a trip to Orlando, Florida. He is buried in Mound Hill Cemetery in Eaton, Ohio, and is honored by Miller Hall at the University of Kentucky.

Sources

A.M. Miller may testify in Scopes' evolution trial. *Lexington Herald*, June 15, 1, 5.
Miller, Arthur M. 1925. What price fundamentalism? *Lexington Herald*, June 28, 6.
Several expert witnesses will appear in Dayton. 1925. *Kingsport (TN) Times*, June 26, 1.

Morgan, John Harcourt Alexander 1867–1950

John Harcourt Alexander Morgan was born on August 31, 1867, into a farming family in Kerwood, Canada. In 1919, following a distinguished career as an entomologist and educator, Morgan became the thirteenth president of the University of Tennessee. He soon convinced legislators and Governor Austin Peay to allocate several million dollars to the university, which he used to expand the university and increase its enrollment. In 1923, Morgan fired several "troublemaker" professors, including the eccentric John Randolph Neal, who later became John Scopes' lead counsel at the Scopes Trial.

By 1925, the University of Tennessee had become increasingly reliant on state appropriations, and Morgan was hesitant to challenge or upset legislators who decided his school's funding. When the legislature began debating the Butler Act, Morgan—a founding member of the Tennessee Academy of Science—worried that speaking out against the bill would jeopardize the university's funding. Unlike university presidents in Kentucky and elsewhere,[28] Morgan remained silent as Butler's antievolution bill moved through the Tennessee legislature. He told Governor Peay, "The subject of Evolution so intricately involves religious belief, which the University has no disposition to dictate that the University declines to engage in the controversy" (Morgan to Peay, 1925). Morgan encouraged his faculty, other educators, and organizations such as the Tennessee Academy of Science and the Tennessee Department of Education to remain silent as well, and they did. Morgan's silence helped pass the Butler Act, which was the nation's first law that criminalized the teaching of human evolution. This law was used to prosecute John Scopes.

In 1934, after resigning as president of the University of Tennessee, the much-honored Morgan joined the inaugural board of the Tennessee Valley Authority. He retired in 1948. Morgan died on August 25, 1950, and is buried in Greenwood Cemetery in Knoxville, Tennessee. Morgan is commemorated at the University of Tennessee by Morgan Hall.

Source

Morgan, Harcourt A., to Austin Peay, February 9, 1925. Governor Austin Peay Papers, Tennessee State Library and Archives, Nashville, Tennessee.

Morgan, Luther Howard
1910–1984

Luther Howard "Scrappy" Morgan was born on August 19, 1910, in Dayton, Tennessee, to Berry Luther and Mamie Morgan. Howard attended Rhea Central High School, where he was in John Scopes' general science class.[29]

Clarence Darrow stayed at the Mansion on his first night in Dayton. However, when his wife Ruby arrived, Darrow paid the Morgans $500 to rent their Second Avenue home for "however long the trial lasted" (Hayba, 1985). Before, during, and after the trial, reporters tracked down several of John Scopes' students, including Howard. He liked the attention, telling one reporter, "I lapped it up … all about monkeys and things." When Scopes urged Morgan and other students to testify against him "and help [them] put me in jail," Morgan reluctantly agreed, promising, "You can bet your hootin' nanny they wouldn't get me to squeal on Prof. Scopes another time" (Sumner, 1925).

On the trial's fourth day of proceedings (i.e., Wednesday, July 15), Howard was the second of four witnesses—and the first of two students—called by the prosecution to testify. When asked by Attorney General Tom Stewart if Scopes had ever taught him anything about evolution, Morgan replied, "Yes, sir," adding that Scopes "said that the earth was once a hot molten mass, too hot for plant or animal life to exist upon it. In the sea, the earth cooled off; there was a little germ of a one cell organism formed, and this organism kept evolving until it got to be a pretty good-sized animal, and then it came on to be a land animal, and it kept on evolving, and from this was man." (Defense attorney Arthur Garfield Hays, apparently impressed with Howard's testimony, instructed Howard to "Go to the head of the class.") When Stewart then asked Morgan if Scopes had classified man "with reference to other animals," Morgan said that Scopes taught him that humans are the culmination of evolution, that humans are classified as mammals, and that "man had a reasoning power that [the other mammals] did not."

Darrow then asked Morgan if Scopes had classified "dogs, and horses, monkeys, cows, man, whales—he said all of those were mammals?" Morgan replied, "Yes, sir," but added, "I don't know about the whales." Morgan testified that Scopes had not taught him anything that was wicked, and when Darrow asked if Scopes' teaching "has not hurt you any, has it?" Morgan replied, "No, sir." This ended Morgan's testimony. As Howard was leaving the witness stand, Darrow turned to Precious Rappleyea (trial instigator George Rappleyea's wife) and asked, "Would you believe this is the twentieth century?" (Linder, 2020). Morgan was paid $4 to be available for four days to testify.

When Bryan told Darrow to ask Morgan's parents "how their children's minds are being poisoned," Howard's mother answered, "The teaching of evolution hasn't hurt me or my boy.… I wish I knew more about it." Scopes later asked Morgan if he had "forgotten all I taught you," to which Howard admitted that he'd "told [Darrow] all I could remember, but I was so scared Darrow would ask me a lot of hard questions that I forgot most of it" (Mothers deny Scopes hurt pupils' religion, 1925; State rests, 1925). Morgan impressed several of the reporters covering the trial; as reporter W.O. McGeehan noted in *Harper's Magazine*, "It is not at all unlikely that one of these days young Howard Morgan may be governor of the state of Tennessee or attorney-general.… All the pupils worshiped" Scopes (McGeehan, 1925).

After graduating from Rhea Central High School, Morgan enrolled at William Jennings Bryan University, where he played tennis and basketball. He then worked for several years at Dayton Veneer and Lumber Mills, after which he moved to Anaheim, California. Morgan died on December 7, 1984, and is buried in a family plot in Dayton's Buttram Cemetery. The Morgan's home in Dayton was demolished in 2008.

Sources

Hayba, Betsy Jones. 1985. Locke and Cunnyngham recall Scopes Trial. *Cleveland Daily Banner*, July 28, 38–39.

Linder, Douglas O. 2020. The student witnesses in the Scopes Trial. Accessed October 1, 2020, at https://famous-trials.com/scopesmonkey/2394-the-student-witnesses-in-the-scopes-trial.

McGeehan, W.O. 1925. Why pick on Dayton? *Harper's Magazine*, 151 (October), 623–627.

Mothers deny Scopes hurt pupils' religion. 1925. *Detroit Free Press*, July 19, 1.

State rests on brief testimony against Scopes. 1925. *St. Louis Post-Dispatch*, July 16, 2.

Sumner, Allene. 1925. Pupils of Scopes in "mess" forced to "blubber" on him. *Tampa Times*, July 4, 2.

National Theatre
Est. 1921

National Theatre, known since 1980 as the David T. Nederlander Theatre, is a 1,232-seat Broadway theater at 208 West 41st Street in New York City. Producer Herman

Shumlin (1898–1979), in association with Margo Jones (1911–1955), opened Jerome Lawrence and Robert E. Lee's *Inherit the Wind* at National Theatre on April 21, 1955. The production, led by veteran actors Paul Muni and Ed Begley, and featuring Tony Randall in his first major stage role, received generally good reviews. Shumlin invited John Scopes to attend the play, but Scopes declined the invitation, saying he was "too busy and not interested enough to fly to New York to see it" (Scopes and Presley, 1967).

When the play's 806th (and final) performance ended on June 22, 1957, *Inherit the Wind* was Broadway's most successful and longest-running play.[30] Unlike Stanley Kramer's movie version of *Inherit the Wind* in 1960, the play *Inherit the Wind* was a commercial success.

Source

Scopes, John T., and James Presley. 1967. *Center of the Storm: Memoirs of John T. Scopes*. New York: Holt, Rinehart and Winston, 266.

Neal, John Randolph, Jr.
1876–1959

John Randolph Neal, Jr., was born John O'Brien Neal on September 17, 1876, in Rhea Springs, Tennessee. His father, John Randolph Neal, was an officer in the Confederate Army, and from 1885 to 1889 served in the U.S. House of Representatives. After his father died, John O'Brien Neal took his father's name. Neal's mother, Mary Elizabeth Brown Neal, helped raise John and his four brothers and sisters. Neal's wealthy family owned large amounts of Rhea County, including much of the land now beneath Watts Bar Lake.

Neal entered the University of Tennessee at age 14 and earned an A.B. degree (with honors) in 1893. He then earned an LL.B. and an M.A. degree from Vanderbilt in 1896, and a Ph.D. in History from Columbia University in 1899. Neal then moved to Denver, Colorado, where he taught law at the University of Denver. In 1906, Neal was elected to the Tennessee House of Representatives for the district including Meigs and Rhea counties. When the legislature was not in session, Neal taught at the University of Denver (Hicks, 1969).

In 1908, Neal was elected to the Tennessee state senate. While a legislator, Neal introduced 40 bills, 16 of which passed. These bills established school districts in every county, improved funding for public schools, ensured a permanent yearly allocation for the University of Tennessee, and improved safety standards in mines. Neal was not overly popular, even in his own party, and was ousted from office two years later. In 1909, Neal began teaching at the University of Tennessee in the spring, and at the University of Denver in the fall. Eight years later, Neal secured a full-time job at the University of Tennessee. Neal was popular with students, joined student groups, and instituted an award for oration. In 1919, Neal was considered as president for the University of Tennessee, but the job went to Harcourt Morgan instead. Six years later, as president, Morgan remained silent as the Tennessee legislature passed the Butler Act, which was used to prosecute John Scopes.

Neal was disliked by administrators, who accused him of being late for or missing classes, seldom recording attendance, refusing to grade papers or submit grades on time,

John Randolph Neal (right) was John Scopes' chief counsel. Here, Neal rests with Scopes (center) and trial instigator George Rappleyea. In the background are Rappleyea's car and Cumberland Coal & Iron Company (i.e., the remnants of Dayton Coal & Iron Company) (RCHGS).

and using class time to talk about current events instead of the curriculum. On July 17, 1923, Neal was among seven professors fired by Morgan who, according to one story, was "getting rid of a bunch of atheists." Ninety-five percent of the law students signed a petition urging that Neal be retained, and Neal—who in 1923 had defended a colleague criticized for teaching evolution (Ginger, 1960)—denounced the charges against him as being "glaringly false." Nevertheless, an investigative committee that included Tennessee Governor Austin Peay voted 5 to 2 to uphold Neal's termination. (Peay cast one of the two dissenting votes.) In response, Neal condemned Morgan as morally and mentally incompetent. In 1925, when the University of Tennessee announced that it would no longer give the John R. Neal Award, Neal began giving the award himself (Hicks, 1969).

In 1924, Neal opened the John Randolph Neal School of Law in Knoxville for part-time students. Neal endowed the school with 1,000 acres of land valued at $100,000. Classes always began on September 17 (Neal's birthday), at which time Neal told his students that he was fired from the University of Tennessee because he allowed students to smoke in class. He then distributed cigars and told students to smoke, adding that they could then tell people that they were required to smoke in Neal's law school (Hicks, 1969). For several years, Neal's school produced more law graduates than did the University of Tennessee. However, after more than 40 students graduated in 1935, enrollments began to decline. Only 15 students graduated in 1939, and soon thereafter no one enrolled (although Neal would continue to announce the school's opening). Neal's school closed in 1943, when Tennessee began requiring full-time attendance at law schools.

When Neal heard that John Scopes had been arrested, he—in Scopes' words—"appeared in Dayton and more or less appointed himself my counsel," adding, "Boy, I'm interested in your case and, whether you want me or not, I'm going to be here. I'll be available twenty-four hours a day" (Scopes and Presley, 1967). Although Darrow and Malone received most of the defense team's publicity, Neal was Scopes' chief counsel. From June 6 to 11, Scopes, Neal, and Rappleyea met in New York with scientists (e.g., Henry Fairfield Osborn, Charles Davenport) and other defense attorneys. On June 23, Clarence Darrow and Bainbridge Colby spoke at the graduation ceremony at which 21 students received diplomas from Neal's law school. At that ceremony, Neal—whom Mencken described as "a good lawyer and an honest man"—awarded Darrow and Colby honorary law degrees (de Camp, 1968).

Neal, who wanted Scopes' trial to focus on teachers' rights and the freedom to teach and learn, sometimes disagreed with Darrow, who wanted to expose Bryan's ideas about science and religion. Neal was initially popular with Daytonians, but his popularity faded when, on July 3, he petitioned federal judge John Gore to move Scopes' trial to federal court. Gore rejected Neal's petition. When Darrow arrived in Dayton, he was met by Neal, who showed him around town. For Neal, Scopes' case involved "the most vital issue—human freedom—that has ever arisen in America, transcending the fundamentals underlying the Civil War" (Hicks, 1968, 1969). Before the trial began, Neal warned, "Every instructor who teaches science in the public schools with the present textbooks is guilty" of violating the Butler Act, and that "if the statute is not held unconstitutional by the court, we shall insist on its enforcement by the prosecution. The best way to get rid of a bad law is to enforce the law" (Other teachers equally guilty, 1925).

After the trial, Neal was in charge of Scopes' appeal. Despite four reminders from the court, Neal missed the deadline for filing the bill of exceptions with the Tennessee Supreme Court.[31] This precluded the defense from challenging any of Raulston's rulings during the trial, including those about expert testimony and the opening of proceedings with prayers. The only option left for the defense team was to challenge the validity of the law. After the Tennessee Supreme Court set aside Scopes' conviction, Darrow and the other defense lawyers agreed that Scopes' case was over, but Neal announced, "I do not mean to accept this defeat as final, but will persist…." Despite Neal's claim, Scopes' case was over.

Neal opposed allowing electric companies to build dams on the Tennessee River and its watershed. Throughout the late 1920s, Neal attended numerous meetings about development of the river; Neal, whom the *Knoxville News-Sentinel* described as "unqualified to vote, uninvited to attend," was often the lone dissenter on a variety of votes (Hicks, 1969). By the 1930s, however, public opinion began to side with Neal. When the Tennessee Valley Authority (TVA) was created in 1933, several people recommended that Neal be made a member of its Board of Directors, but that appointment went to Harcourt Morgan instead.

When creating nearby Watts Bar, TVA forced Neal to sell some of his land in Rhea Springs for $55,000 (Hicks, 1968, 1969). As was his custom with most of the checks that he received, the wealthy Neal did not cash the $55,000 check, instead keeping it for several months folded in his pocket with dozens of other uncashed checks. Neal often claimed to represent "the people," for which he paid all of his own expenses. Throughout his career, Neal claimed credit for some things that he did not do, but he failed to get credit for many other things that he did accomplish.

Neal was a perennial candidate for public office. Indeed, between 1924 and 1954, he ran for the U.S. House of Representatives once, for governor nine times, and for the U.S. Senate 18 times. During his 1912 run for governor, Neal taught at the University of Denver during his campaign, and in 1946 he ran for governor and senator simultaneously. In several of his runs for governor, Neal promised, if elected, to repeal the Butler Act. He never came close to winning. During this time he also defended striking workers, murder suspects, and mistreated prisoners, all *pro bono*.

Neal's legal skills were often overshadowed by his eccentricity, and especially by his disregard for personal appearance and hygiene. He wore the same clothes for days at a time, rarely bathed, smelled bad; and was often mistaken as a bum; as one colleague bluntly noted, "he was filthy." After staying in a Knoxville hotel for several days, where he filled his room's bathtub with books, Neal was forced out because of his hygiene. One writer described him "as dirty as some early Christian saints," while others joked, "Who does Judge Neal get to wear his shirts for a week before he puts them on?" (de Camp, 1968). Scopes admitted that Neal "revulsed" most people, and a newspaper claimed that Neal looked "as if he had just come from beneath a boxcar," a look that Neal claimed was his way of protesting conformity (Hicks, 1969, 1970). Scopes also praised Neal as "one of the warmest-hearted men I have ever known…. He was a good man and a clear thinker, a much rarer person than most spotless, clean-shaven, and shorn clotheshorses I have seen … [Neal] played as important a role as anyone in my defense" (Scopes and Presley, 1967).

Neal's last campaigns were in 1954, at which time he simultaneously lost races for the senate (to Estes Kefauver) and governor (to Frank Clement). Neal spent his last years at his home in Spring City, Tennessee. During the summers, he would often be seen in a dirty white linen suit walking between his home and Dayton.

The 83-year-old Neal died of pneumonia in Rockford, Tennessee, on November 23, 1959. He is buried in a family plot in Ault Cemetery in Postoak, Tennessee.

Sources

de Camp, L. Sprague. 1968. *The Great Monkey Trial.* Garden City, NY: Doubleday.
Ginger, Ray. 1960. *Six Days or Forever? Tennessee v. John Thomas Scopes.* London: Oxford University Press, 44.
Hicks, Bobby E. 1968. The Great Objector: The public career of Dr. John R. Neal. Master's Thesis, University of Tennessee.
______. 1969. The Great Objector: The life and career of Dr. John Randolph Neal. *East Tennessee Historical Society* 41, 33–36.
______. 1970. Dr. John R. Neal: The man, the myth. *Rhea Area News*, July 7, 3.
Howell, Alice, and Lucile Deaderick (Eds.). 1976. *Heart of the Valley: A History of Knoxville, Tennessee.* Knoxville: East Tennessee Historical Society, 561–563.
Other teachers equally guilty as Scopes—Neal. 1925. *Chattanooga News*, May 18, 1.
Scopes, John T., and James Presley. 1967. *Center of the Storm: Memoirs of John T. Scopes.* New York: Holt, Rinehart, and Winston, 63.

Norris, John Franklyn
1877–1952

John Franklyn "Frank" Norris was born on September 18, 1877, in Dadeville, Alabama, to a devout mother and an alcoholic sharecropper father. When 15-year-old Frank was shot by horse thieves, his mother nurtured him back to health and instilled in him a devout religious faith. After his family moved to central Texas in the 1880s, Frank became pastor at Mount Antioch Baptist Church in Mt. Calm, Texas, where the church's membership grew from 13 to almost 1,000 in three years. He then enrolled at Baylor University, and later earned a Master of Theology degree (as valedictorian) in 1905 from Southern Baptist Theological Seminary in Louisville, Kentucky. After a two-year pastorship at McKinney Avenue Baptist Church in Dallas, Norris edited the *Baptist Standard* for three years. In 1909, Norris sold his interest in the *Baptist Standard* and became the pastor of Fort Worth's First Baptist Church, where he served for 44 years until his death. Although Norris tolerated no dissent and often told disgruntled members to "go to hell," membership in his "Church of the Cattle Kings" grew from 150 to more than 12,000 members. After Sinclair Lewis "satisfied a desire of a great many years standing [by going] to hear Dr. J. Frank Norris," he told reporters that he admired Norris's "eloquence and vigor," adding, "I have never seen before so many people at church at once" (Entzminger, 1970; Sears, 1937).

Norris had a seemingly unquenchable thirst for publicity, controversy, and fame. He published KKK advertisements in his newspapers, sponsored KKK shows in his church, and endorsed Klansmen in elections. At his circus-like church services, Norris hosted delegations of monkeys when he mocked and condemned evolution, violently threw biology textbooks—"$24.50 worth of them"—into trash cans when he denounced evolution, smashed 50 bottles of moonshine when he condemned alcohol, and had a horse ridden into the sanctuary when he baptized a prize-winning cowboy (J. Frank Norris, 2020). Norris was repeatedly arrested for felonies such as perjury, arson, and murder (in his church office), all of which resulted in acquittals and growth of his church. By the early 1920s, his rebuilt church's sanctuary seated more than 5,200 people and was usually overflowing when Norris preached (Russell, 1972). Norris, a founding member of the World's Christian Fundamentals Association (WCFA), also began one of the nation's first radio ministries (station KFQB, standing for "Keep Folks Quoting the Bible"). Although Norris regularly hosted other antievolution crusaders such as Billy Sunday, William

Bell Riley, and William Jennings Bryan, he criticized Southern Baptist leaders as "old baboons," "lepers," "weeds," "worthless bastards," and, in one instance, "a beer guzzling bunch of hair," after which he and his congregation were kicked out of every Baptist organization in America (Perjury, murder trials, 1952; Russell, 1972). Many preachers denounced Norris, with one even claiming that he would "see Frank Norris frying in the bottomless pits of Hell." Norris, a classic militant fundamentalist who brought fundamentalism to the South, created many of today's unflattering caricatures of fundamentalists as bigoted, ignorant Southerners (Entzminger, 1970).

Norris was at the forefront of the fundamentalists' antievolution sentiments touted throughout the Scopes Trial. To Norris, who equated evolution with "the poisonous gas of German armies … [that is] sweeping through our schools," "if you deny the Genesis account of creation, you had just as well throw the Bible in the fire" (Hankins, 1996). When Norris hosted the 1923 meeting of the WCFA, he staged mock trials of evolution teachers; with fellow-fundamentalist William Bell Riley presiding, all of the teachers were convicted of teaching evolution and denounced by Norris as "damnable and despicable" because they had "hands dripping with innocent blood" (Hankins, 1996). Like other fundamentalists, Norris believed that college students were losing their Christian faith after hearing about "that hell-born, Bible-destroying, deity-of-Christ denying, German rationalization known as evolution" from "some beer-guzzling German Professor of Rationalism." In emotional sermons in which he roamed the stage while weeping and shouting, Norris claimed, "I would rather my boy be in heaven without knowing one thing about bugs, lizards, apes, or chimpanzees than to be in hell with a basket full of degrees and a post-graduate [degree] in the sciences on earth" (Norris, 1921). Because of Norris, eight "anthropoid ape" professors at nearby SMU and Baylor University (Norris's alma mater) were fired for allegedly teaching "animal ancestry." After renting an auditorium near Baylor, Norris promised to "hang the apes and monkeys" who were teaching evolution there (Entzminger, 1970).

There was little that Norris would not do to attract attention, including promising that God would inflict accidents, poverty, and venereal disease on his critics. When a lawyer who had prosecuted him for arson died in a car crash, Norris brought a whiskey bottle from the crash scene to church "with a lobe of brains in it," after which he reminded his congregation that "the wages of sin is death" (Entzminger, 1970; Ledbetter, 1973). Norris's antics drove hundreds of members away (600 in 1911 alone), but they were quickly replaced by thousands of others wanting to be part of what a local newspaper called "the best show in town." Some people believed that Norris "saved the Southern Baptists from evolution," but the *Atlanta Constitution* described him as "one good, sound reason why there are 50,000,000 Americans who do not belong to any church" (McGill, 1947).

As the Scopes Trial approached, Norris promised Bryan that he would be in Dayton to "fight for the Bible against the onslaught of the infidelity of Clarence Darrow," but he never showed up. Instead, Norris sent L.V. Evridge, the managing editor of Norris's popular *The Searchlight* newspaper. Norris later sold Evridge's report as the "Only Authentic Book on Dayton Trial." When he read reports of Bryan's poor showing against Darrow, Norris cited Evridge's book to denounce the press for misreporting Bryan's testimony.

On the day after Bryan died, Norris received a letter that Bryan had written from Chattanooga on the day before he died: "My Dear Mr. Norris, Well, we won our case. It woke up the country if I can judge from the letters and telegrams. Am just having my speech (prepared but not delivered) put into pamphlet form. Will send you a copy. I think it is the strongest indictment of evolution I have made. Much obliged to you for your part in getting me in the case. Much obliged to you for Evridge. He is delightful and very efficient. I wish you would let me correct my part in the trial before you publish it. Sorry you were not there. Yours, Bryan" (Bryan to Norris, 1925; Stokes, 2011). A week later, Norris led a memorial service for Bryan in Fort Worth, telling the 6,000 mourners that he could hear Bryan in heaven saying that "the last act of my life was to stand yonder at Dayton, Tennessee, against the infidelity of the world for a gospel that can make a heaven like this" (Stokes, 2011).

In 1926, Norris accused Fort Worth's mayor H.C. Meacham of improperly funneling money to help Catholics, adding that Meacham wasn't "fit to manage a hog-pen." When Norris was visited in his church office by Meacham's supporter Dexter Chipps, Norris pulled a pistol from his desk and shot the unarmed Chipps several times, killing him. Norris then called his wife and the police, telling them, "I just killed me a man" (Taylor, 2016). As Norris faced the death penalty, his many supporters jumped to his defense; in New York, antievolution crusader John Roach Straton pledged to support Norris "no matter what," and the KKK announced it was "ready to lend whatever assistance" it could. Norris claimed that he had acted in self-defense because Chipps "was reaching for a gun." The jury, after deliberating for only 40 minutes, acquitted Norris, after which he told his congregation, "I offer no apology for what I have done" (Norris found not guilty, 1927; Entzminger, 1970). As one newspaper in Fort Worth noted, "the 11th Commandment is 'Thou shalt not mess with J. Frank Norris'" (Stokes, 2011).

In 1935, Norris accepted the pastorate of Temple Baptist Church in Detroit, at one point conducting a baptismal service along the Detroit River attended by 40,000 followers. This made Norris, who travelled weekly between Detroit and Fort Worth, the pastor of the largest Protestant congregation in North America (Russell, 1972). By 1946, the combined membership of Norris's two churches exceeded 26,000 members. Today, a large statue of Norris welcomes visitors to Arlington Baptist College's campus in Ft. Worth.

By the end of his life, Norris—who described himself as "The Texas Cyclone"—had been deserted by many of his friends and family. In 1950, Temple Baptist Church voted 3,000 to 7 to fire Norris. His friend Benjamin Bogard (1868–1951), a fundamentalist preacher and Klansman who had been instrumental in Arkansas's banning the teaching of human evolution,[32] claimed, "When I get to heaven, I expect to find Frank Norris there in spite of the wicked streak that runs through him" (Bogard testifies, 1944).

Norris, the South's leading fundamentalist, died of a heart attack at a youth camp at Keystone Heights, Florida, on August 20, 1952. Following a funeral attended by 5,000 people and broadcast from his church, a miles-long, 18-policecar procession took Norris's body to Fort Worth's Greenwood Cemetery, where he was buried about 400 yards northeast of Chipps. Norris rests beneath the inscription "American's Foremost Fundamentalist." Countless subsequent fundamentalists, such as Jerry Falwell, often proudly told their congregations, "I was trained by men who were trained by J. Frank Norris" (Taylor, 2016).

Sources

Bogard testifies against Norris. 1944. *The Bible Banner*, 6 (13) (July–August), 18b.

Bryan, William Jennings, to Frank Norris. n.d. (July 23–26). Arlington Baptist College Museum and Archives, Ft. Worth, Texas.

Crow, Kelvin. 1997. Militants, moderates, and millionaires. *Christian History*, 16 (3), 28–31.
Entzminger, Louis. 1970. *The J. Frank Norris I Have Known for 34 Years*. Fort Worth: New Testament Books, 3, 109, 112, 164–168, 177.
Hankins, Barry. 1996. *God's Rascal: J. Frank Norris and the Beginnings of Southern Fundamentalism*. Lexington: University Press of Kentucky.
J. Frank Norris: One foot in the pulpit, one foot in the witness stand. 2020. Accessed November 4, 2020, at https://hometownbyhandlebar.com/?p=2015&cpage=4.
Ledbetter, Patsy. 1973. Defense of the faith: J. Frank Norris and Texas fundamentalism, 1920–1929. *Arizona and the West*, 15 (1), 45–62.
McGill, Ralph. 1947. J. Frank Norris gets shouted down. *Atlanta Constitution*, May 8, 10.
Norris found not guilty, weeps aloud. 1927. *Fort Worth Record-Telegram*, January 26, 1.
Norris, J. Frank. 1921. The inspiration of the Scriptures. *Searchlight*, May 21, 1.
Perjury, murder trials marked pastor's career. 1952. *Fort Worth Star-Telegram*, August 21, 1–2.
Russell, C. Allyn. 1972. J. Frank Norris: Violent fundamentalist. *The Southwestern Historical Quarterly*, 75 (3), 271–302.
Sears, Mary. 1937. Sinclair Lewis, here for lecture. *Ft. Worth Star-Telegram*, November 1, 12.
Stokes, David R. 2011. *The Shooting Salvationist: J. Frank Norris and the Murder Trial that Captivated America*. Westminster, MD: Steerforth.
Taylor, Justin. 2016. Goodbye Mr. Chipps: The day a fundamentalist pastor shot and killed a Catholic layman in his Baptist church. July 18. Accessed November 16, 2020, at https://www.thegospelcoalition.org/blogs/evangelical-history/90-years-ago-today-when-a-pastor-in-texas-shot-and-killed-mr-chipps-in-his-baptist-church/.

Osborn, Henry Fairfield 1857–1935

Henry Fairfield Osborn was born on August 8, 1857, into a rich, prominent family in Fairfield, Connecticut. From 1873–1877, Osborn attended Princeton University, earning a B.A. in geology and archaeology while being mentored by famed paleontologist Edward Drinker Cope. After studying anatomy with Thomas Huxley in Europe, Osborn returned to Princeton, where he earned a Sc.D. in paleontology in 1880. In 1891, Osborn became a zoology professor at Columbia University and a curator at the American Museum of Natural History (AMNH). Osborn became president of the museum's Board of Trustees in 1908, a position he held until 1933. Under Osborn's leadership, AMNH became one of the world's premier natural history museums and advocates of evolution. Osborn described several famous dinosaurs, including *Tyrannosaurus rex* (1905), *Albertosaurus* (1905), and *Velociraptor* (1924). One of Osborn's students, Arthur Miller, taught John Scopes geology at the University of Kentucky.

Osborn wanted to remove apes from humans' family tree, for he believed "not that man had descended from monkeys or from the apes, but that he has a long and independent line of ascent of his own," and that humans' ancestor, when found, "will not be an ape, but it will be surprisingly pro-human" (Lewin, 2005).[33] Central Asia was Osborn's Garden of Eden; he dismissed the 1924 discovery of *Australopithecus africanus* in Africa as well as claims that humans originated there, and used AMNH to fund expeditions to Asia to search for human ancestors. Osborn based his "Dawn Man Theory" of human origins on Java Man, Neanderthal, his Jefferson-like Cro-Magnon, and especially the controversial Piltdown Man (*Eoanthropus dawsoni*), which allegedly showed that England was once home to prehistoric humans and that Darwin's theory of natural selection was wrong.[34] Osborn, who claimed that there was "irrefutable evidence that man-apes wandered over from Asia into North America," never relinquished his purposeful evolution or his so-called (and allegedly more dignified) "Out of Asia" theory, dogmatically proclaiming that "no existing form of anthropoid ape is even remotely related to the stock which has given rise to man" (Osborn states case, 1925). AMNH provided materials promoting Osborn's claims to virtually all textbook publishers, thereby ensuring that Osborn's claims remained in textbooks well into the 1970s (despite the fact that Piltdown Man was exposed as a fraud in 1953). Osborn also claimed that Caucasians' evolutionary struggle had endowed a primitive quality that would be ruined by racial interbreeding. In Osborn's AMNH, racist and eugenic themes were prominent.

Osborn, who equated human evolution with a struggle for spiritual salvation, helped John Scopes' attorneys recruit scientists as expert witnesses for the Scopes Trial (Regal, 2002). Like the scientists who came to Dayton, Osborn reassured believers, "God did use evolution as His plan," and an AMNH movie told viewers, "Some call it evolution, and others call it God" (Osborn, 1925). Many people found comfort in Osborn's claim that humans did not share an ancestry with African apes.

In June 1925, Scopes met with Osborn, Malone, Darrow, and others at AMNH, at which time Osborn gave Scopes money to help him pay for unexpected expenses (Regal, 2002). Osborn, with fellow-eugenicists Edwin Grant Conklin and Charles Davenport, was a member of the AAAS's "Committee on Evolution" that defended the teaching of evolution in public schools. Although he was opposed to Darrow leading Scopes' defense, Osborn urged Scopes' attorneys to make "the fact that evolution and religion are not in any way inconsistent the backbone of the defense" (Regal, 2002). The defense team, which sought Osborn's committee's advice before and during the trial, relied heavily on Osborn; as John Randolph Neal later noted, "Professor Osborn possibly did more than any other scientist to aid us in the Scopes case" ("Battle of Dayton" on again, 1927).

Despite his importance to Scopes' defense, Osborn apparently cared little about the civil liberties associated with Scopes' trial, claiming instead that "William Jennings Bryan is the man on trial; John Thomas Scopes is not the man on trial.... [P]ersonal rights, ... rights of free speech, constitutional rights, educational liberty ... are merely temporary side issues and will fade into insignificance.... If Scopes has been teaching the truth to his students, he will win; if he has been teaching untruths, he will lose and deserve to lose" (Osborn, 1925).

As the Scopes Trial approached, the ACLU promised that Osborn would be in Dayton to support Scopes (Many noted scientists, 1925). Although Scopes' upcoming trial seemed like an ideal venue for Osborn to confront Bryan, Osborn—who was "second only to Albert Einstein as the most popular and well-known scientist in America"—never showed up. Osborn claimed that his wife was too sick for him to travel, yet he traveled across the country near the same time for other reasons. Osborn, a virulent eugenicist and the first American to receive an honorary degree from Adolf Hitler, may have been uncomfortable with Darrow's opposition to eugenics (Davis, 2005).

Osborn had often ridiculed Bryan's opposition to the teaching of evolution. For example, in his *The Earth Speaks to Bryan*, which he dedicated to Scopes, Osborn used a tooth discovered in 1922 in Nebraska to promote *Hesperopithecus haroldcookii*, which Osborn proclaimed was "the first anthropoid primate found in North America" (Osborn, 1922). To Osborn, the tooth was a "still small voice" through which "the Earth spoke to Bryan [about the truth of evolution] ... from his own native state of Nebraska ... but he fails to hear a single sound" (Osborn, 1925). Osborn rushed *The Earth Speaks to Bryan* into print and sent copies to dignitaries in Dayton before the trial. However, a growing number of scientists were questioning Osborn's claims about the tooth. When Bryan heard of these criticisms, he ridiculed Osborn's "grotesque gullibility," adding that "a man who can do this [i.e., reconstruct a new species from a worn tooth] could take a keyhole and build a house around it" (Bryan, 1921). Bryan was later vindicated; the tooth on which Osborn had based a new species of primate was a premolar tooth from an extinct, pig-like peccary (*Prosthennops serus*; Gregory, 1927). Despite his many years of attacking and ridiculing Bryan, Osborn ducked the chance to confront Bryan in Dayton.

Osborn died in his home on November 6, 1935, in Garrison, New York, where he is buried by his wife Lucretia in Saint Philip's Church Cemetery. Osborn's once prominent bust and memorial in AMNH have been moved to a seldom-visited part of the museum.

Sources

"Battle of Dayton" on again, says Neal. 1927. *New York Times*, September 12, 14.

Bryan, William Jennings. 1921. *The Bible and Its Enemies. An Address Delivered at the Moody Bible Institute of Chicago*. Chicago: The Bible Institute Colportage Assn., 22.

Davis, E.B. 2005. Science and religious fundamentalism in the 1920s. *American Scientist*, 93, 253–260.

Gregory, W.K. 1927. *Hesperopithecus* apparently not an ape nor a man. *Science*, 66 (1720), 579–581.

Lewin, Roger. 2005. *Human Evolution: An Illustrated Introduction*. Hoboken: Wiley-Blackwell, 56.

Many noted scientists enlisted for defense in evolution trial. 1925. *St. Paul (MN) Press*, June 25.

Osborn, Henry Fairfield. 1915. *Men of the Old Stone Age*. New York: Charles Scribner's Sons.

______. 1922. *Hesperopithecus*, the first anthropoid primate found in North America. *Science*, 55 (1427), 463–465.

______. 1925. *The Earth Speaks to Bryan*. New York: Charles Scribner's Sons, 13, 65.

______. 1925. Osborn states the case for evolution. *New York Times*, July 12, 21.

Regal, Brian. 2002. *Henry Fairfield Osborn: Race and the Search for the Origins of Man*. Farnham, UK: Ashgate.

Paducah, Kentucky
Est. 1830

Paducah, Kentucky, was settled as "Pekin" by Native Americans and European-Americans around 1821, laid out by explorer and surveyor William Clark (1770–1838) in 1827, established by Kentucky in 1830, and incorporated in 1838. Clark, the coleader of the famed Lewis and Clark expedition of 1804–1806, renamed the town "Paducah" for a large, distinctive group of Chickasaw Native Americans who were known to regional settlers as the "Padoucas."

Paducah, which is near the confluence of the Ohio and Tennessee Rivers (about halfway between St. Louis, Missouri, and Nashville, Tennessee), soon became a bustling riverport. The first railroad reached Paducah in 1855, after which the area became a hub for the Illinois Central Railroad connecting Chicago and St. Louis with New Orleans. Before long, Paducah enjoyed a growing "river and rail" economy (Holland, 1994). By 1883, newly arrived English immigrant Thomas Scopes—John Scopes' father—was working as a machinist for the Illinois Central Railroad.

By 1900, Thomas, his wife Mary Alva Scopes, and their children were again in Paducah, where Thomas worked in the railyards (Appendix 1). In subsequent years, several of Thomas and Mary Alva's children also lived there, where they worked in industries ranging from candy-making to meat packing. Thomas and Mary Alva's son John, who was born in 1900 in Paducah, described Paducah as "an Ohio River town that was a symbol of the live-and-let-live philosophy that I was taught at home" (Scopes and Presley, 1967). John visited his parents often, and he stayed with them while he was unemployed. In 1932, John and his wife Mildred Scopes' oldest son, John Jr., was born at Thomas and Mary Alva's home in Paducah.

In 1981, Paducah, Kentucky, commemorated the grave of John Scopes with this roadside marker outside Oak Grove Cemetery (Susan Brooks).

In early 1937, more than 90 percent of Paducah was inundated by "the worst flood in the history of the Ohio Valley" (Welky, 2011). Although Thomas Scopes was now blind, the Scopes clan worked with the American Red Cross to help some of the 27,000 displaced residents who were seeking higher ground. The Scopes' homes just outside of town atop a hill along Berger Road initially hosted 18 guests, but occupancy increased to 45 during parts of the three weeks of flooding. Today, 20 blocks of Paducah's downtown comprise a historic district listed on the National Register of Historic Places. Several downtown buildings display plaques marking the flood's high-water mark, and a floodwall built downtown after the flood features murals of Paducah's history.

In 2020, Paducah had a population of 27,137. Its historic Oak Grove Cemetery (est. 1847) is the resting place of Thomas and Mary Alva Scopes and most of their family, including John and his wife Mildred. A Kentucky Historical Society and Kentucky Department of Transportation Historical Marker (#1698) along Park Avenue near the cemetery's entrance commemorates John Scopes' grave and his famous trial. One side of the marker reads as follows: "GRAVE OF JOHN T. SCOPES Here is buried the man who, at age 24, taught Darwin's theory of evolution to a Dayton, Tennessee, biology class. The Paducah native and University of Kentucky graduate violated a Tennessee law forbidding the teaching of evolution. This test case, tried in Dayton, gained international attention. Popular play, 'Inherit the Wind,' is based on the famous Scopes trial. Over." The opposite side of the marker reads as follows: "SCOPES 'MONKEY TRIAL' The July 1925 trial of John T. Scopes had at issue academic freedom, separation of church and state, and reconciling of science and religion. Scopes' defender was Clarence Darrow; his prosecutor, William Jennings Bryan. Scopes was convicted and fined $100. Later worked as geologist in S. America and La. Law he violated repealed in 1967. Scopes died, 1970. Over."

Sources

Holland, Richard. 1994. *Paducah: Portrait of a River Town*. Paducah, KY: Image Graphics.
Scopes, John T., and James Presley. 1967. *Center of the Storm: Memoirs of John T. Scopes*. New York: Holt, Rinehart and Winston, 5.
Welky, David. 2011. *The Thousand-Year Flood: The Ohio-Mississippi Disaster of 1937.* Chicago: University of Chicago Press, 101.

Peay, Austin
1876–1927

Austin Peay IV was born on June 1, 1876, into a farming family in Hopkinsville, Kentucky. Some sources list his name as "Austin Leavell Peay" and "Austin L. Peay," but he stopped using a middle name by the time he became Tennessee's governor. In 2013, the PBS television show "Genealogy Roadshow" claimed that Peay may have fathered a child with his family's Black housekeeper when he was 14 years old (Smith, 2013).

In 1895, after attending Washington and Lee University (Lexington, Virginia) and Centre College (Danville, Kentucky), Peay began practicing law in Clarksville, Tennessee. Five years later, when he was only 24 years old, Peay was elected to Tennessee's House of Representatives. He accomplished relatively little in his first term, but by 1905 he chaired the Tennessee Democratic Committee. After quitting politics and returning to Clarksville to practice law, Peay ran for governor in 1918, but he was defeated in the Democrat primary by Albert Roberts, who later won the general election. In 1922, Peay ran again for governor, beating Alfred Taylor by a vote of 141,002 to 102,586. As governor, Peay overhauled Tennessee's tax code, created Tennessee's first state parks, improved schools and roads, and transformed a $3 million debt into a $1.2 million surplus. Peay was especially interested in improving agriculture and rural life in Tennessee.

On March 21, 1925, Peay—after getting "tens of thousands of letters" about Butler's bill—signed into law the Butler Act, which banned the teaching of human evolution in Tennessee's public schools. Peay claimed that the law would improve public schools because it "does not require any particular theory of interpretation of the Bible regarding man's creation to be taught in the public schools.... The widest latitude of interpretation will remain as to the time and manner of God's process in his creation of man" (Goetz, 1997). William Jennings Bryan told Peay that Tennessee parents owed Peay "a debt of gratitude for saving their children from the poisonous influence of an unproven hypothesis." On June 15, Rhea County School Superintendent Walter White offered similar praise and, with the prosecution, asked Peay to come to Dayton to testify in Scopes' trial. Peay responded, "I know no fact which could conceivably be useful to either side in that suit as a witness. That case should be tried in an hour. It is about as simple a proposition as could be stated, and the great hurrah about it is unnecessary and unfortunate. The statute, I have no doubt, is entirely constitutional. We have the right to say through our law-making body, that the faith and religion of our children shall not be destroyed by teachers who poison their minds with stuff that no science has established and which belongs in no reputable textbook. Tennessee needs no sympathy nor commiseration. Our state has taken a great and forward position which deserves and will certainly receive in due time the admiration and appreciation of the Christian world. I have a profound contempt for those who are throwing slurs at Tennessee for having this law. In my judgment, any state had better dispense with its schools than with the Bible. We are keeping both" (Gov. Peay's letter to Walter White, 1925). Peay did not attend the Scopes Trial.

On November 5, 1926, Peay overturned the first shovel of dirt atop a hill in Dayton where William Jennings Bryan University would be built. During his speech that followed, Peay told more than 1,000 onlookers that Tennessee needed more Christian teachers so that "there would be no conflict between science and religion in education."

On November 5, 1926, Tennessee governor Austin Peay overturned the first shovel of dirt at the groundbreaking ceremony for William Jennings Bryan University in Dayton. Peay promised thousands of onlookers that the university would "bear a magic name.... Fortunate will be your university to bear the name of Bryan" (RCHGS).

According to Peay, "Fortunate will be your university to bear the name of Bryan" (Livesay, 2020).

In 1926, after defeating Rhea County School Superintendent Walter White to win a third term as governor, Peay's health declined. He died in the Governor's Residence from a cerebral hemorrhage on the evening of October 2, 1927, and was buried in Greenwood Cemetery in Clarksville, Tennessee. Before he died, Tennessee honored Peay by creating Austin Peay Normal School in Clarksville, Tennessee. The school's name was later changed to Austin Peay State University, and their athletic teams are "The Governors." As of 2022, Peay remains the only governor of Tennessee to die in office.

Sources

Goetz, David. 1997. The monkey trial. *Christian History*, 16 (3), 10–18.
Gov. Peay's letter to Walter White. 1925. *Nashville Banner*, June 26, 1.
Livesay, Corinne. 2020. *William Jennings Bryan: A Difference Maker*. Dayton, TN: Bryan College.
Message from the Governor. 1925. *Journal of the House of Representatives of Tennessee*, Regular Session (March 23), 741–745.
Smith, Chris. 2013. PBS show claims Gov. Austin Peay was father of black housekeeper's child. Clarksville Now. Accessed October 18, 2020, at https://clarksvillenow.com/local/pbs-show-claims-gov-austin-peay-was-father-of-black-housekeepers-child/.

Potter, Charles Francis 1885–1962

Charles Francis Potter was born on October 28, 1885, in Marlboro, Massachusetts. He became an ordained minister at age 17 and later earned degrees from Bucknell University (1907, 1916) and Newton Theological Institution (1916). Potter began his career as a Baptist preacher, but his growing theological liberalism prompted him to convert to Unitarianism in 1914. In 1920, after serving several congregations, Potter became pastor of New York City's West Side Unitarian Church.

In 1923–1924, Potter and New York fundamentalist preacher John Roach Straton faced off in four highly publicized debates that were broadcast and reported nationwide. During each of the three debates held at Carnegie Hall in front of more than 3,000 spectators, Potter warned of "the ominous challenge of democratic education by fanatic religion" (Straton, 1988). Although Potter lost the evolution debate by unanimous decision, the nationwide publicity associated with the debate focused the public's attention on the controversy.

On April 27, 1924, just days before his final debate with Straton, Potter hosted "Evolution Day" at his church, which included a musical rendition of William Herbert Caruth's poem, "Each in His Own Tongue" ("Some call it evolution/And others call it God"). During the service, Potter unveiled a bronze sculpture by Carl Akeley named "Chrysalis" depicting a man emerging from the skin of a gorilla. Although Straton condemned Potter's service as desecration of the church, Potter continued to preach about the dangers of fundamentalism and Bryan, to which Straton responded with sermons about the moral ruin produced by evolution and Darwin. Potter believed that "it is not only the people of Tennessee who should be ashamed. We are all to blame. The educated and enlightened American ought not rest until the possibility of such dense mental darkness is removed" (Evolution holds center stage, 1925). To Potter, the difference between fundamentalism and modernism was "the difference between ignorance and education," and Tennessee's fundamentalists embraced their religion because "they haven't had a chance to find out anything else" (U.S. has bad case of intolerance, 1925; Dayton pastor quits, 1925).

In early July 1925, Potter came to Dayton with defense attorneys Dudley Field Malone and Arthur Garfield Hays to help defend John Scopes. Potter, who believed that "people in the South don't know what evolution is," brought reference books and gave the defense the benefit of his many debates with Straton. In Dayton, Potter met his friend L.M. Birkhead, the pastor of All Souls Unitarian Church of Kansas City, Missouri. Birkhead's wife, Agnes, served as the unpaid stenographer for Darrow and others on the defense team.

Soon after Potter got to Dayton, the Rev. Howard Byrd asked him to preach about evolution at First Methodist Episcopal Church (North). Earlier, when Potter had asked, "What does it matter in real life whether you believe in the Virgin Birth?" many in Dayton branded him an infidel. Now, after learning that Potter would speak in their church, Byrd's outraged congregation promised to "wreck the church" if Potter spoke there. Potter posted a sign at Robinson's Drug Store announcing that "Dr. Potter has cancelled his evolution sermon this morning at the Methodist Episcopal Church owing to the threat of the congregation." Byrd then announced, "I have quit [as pastor]. I have not resigned; I have quit," and moved to Alabama (Dayton pastor quits, 1925). Instead of speaking about evolution at the church that day, Potter spoke about brotherly love on the courthouse lawn.

During the trial, Potter joined Rabbi Jerome Mark (from Knoxville), Congregational minister Fred Hagan (from West Virginia), and Unitarian minister D.M. Welch (from Knoxville) in a petition telling Judge Raulston that "there are many to whom the prayers of the Fundamentalists are not spiritually uplifting and are occasionally offensive," and asking him to "occasionally hear a prayer which requires no mental reservations on our part and in which we can conscientiously participate." On the trial's fourth day (i.e., July 15), Raulston called on Potter for the opening prayer, in which Potter asked God for "the progress of mankind toward Thy truth." Potter's mild prayer was the trial's shortest prayer yet. Potter enjoyed his stay in Dayton, but was surprised when, at Rappleyea's office, a secretary refused to take dictation when she learned that he was a Unitarian minister.

Potter—who dismissed Malone as a "clown of the trial and a Roman Catholic stooge"—predicted a nationwide flood of antievolution laws (de Camp, 1968). The South remained especially receptive to fundamentalism, but even there, most antievolution bills failed, and the flood of such laws predicted by Potter never materialized. (Only two more states—Mississippi and Arkansas—passed laws criminalizing the teaching of human evolution.) Potter, who claimed, "The Scopes Trial signed [Bryan's] death warrant," used the research he had done for Darrow to produce *Is that in the Bible?* (1933), one of his bestselling books. Potter, who understood Rappleyea's role in starting the Scopes Trial, urged Dayton to erect a monument to Rappleyea, but that never happened (Potter, 1951).

Soon after the trial, Potter—feeling that even Unitarianism was too restrictive—resigned his job. By 1927, however, Potter had returned to the ministry at the Church of the Divine Paternity, a Universalist congregation on Manhattan's Upper West Side. In 1929, Potter—frustrated that the church remained too conservative for his liking—resigned and founded the First Humanist Society of New York. Later that year, he performed the first Humanist marriage, which included no mention of God. By now, Potter "had given up my fast dwindling belief in the deity of Jesus and the doctrine of the Trinity.... I was leaving not only Christianity—if Unitarianism is Christianity—but Theism as well." In 1933, Potter was among the original 34 people to sign the first *Humanist Manifesto*, which Potter said would be "a religion of common sense." To Potter, "Man has waited too long for God to do what man ought to do himself" (Charles Potter, 1962). Potter often campaigned against the death penalty and supported women's rights and birth control.

Potter wrote several books, including *The Story of Religion* (1929), *Humanism: A New Religion* (with his wife, Clara, in 1930), and *Humanizing Religion* (1933). He founded the Euthanasia Society of America in 1938, and in 1958 he published his most popular book, *The Lost Years of Jesus*. Over the span of his career, Potter changed from an evangelical Baptist to a radical Humanist.

Potter spent his final years lecturing, writing, and living in New York's Ansonia Hotel. The 76-year-old Potter died of stomach cancer on October 4, 1962, in New York's Polyclinic Hospital. He is buried in a family plot in Maplewood Cemetery in Marlborough, Massachusetts.

Sources

Charles Potter, clergyman, dead—Retired Unitarian minister began Humanist Society. 1962. *New York Times*, October 5, 33.
Dayton pastor quits when congregation bars modernist. 1925. *Des Moines Register*, July 13, 1–2.
de Camp, L. Sprague. 1968. *The Great Monkey Trial*. New York: Doubleday, 152.
Evolution holds center stage in many discussions. 1925. *Hamilton (OH) Journal*, June 15, 1.
Potter, Charles Francis. 1933. *Is that in the Bible?* Garden City, NY: Doubleday.
______. 1951. *The Preacher and I: An Autobiography*. New York: Crown.
______, and Clara Cook Potter. 1930. *Humanism: A New Religion*. New York: Simon & Schuster.
Straton, John Roach. 1988. *Fundamentalist versus Modernist: The Debates Between John Roach Straton and Charles Francis Potter*. New York: Garland.
U.S. has bad case of intolerance. 1925. *Spokane (WA) Spokesman-Review*, October 28, 3.

Prayer

Prayer was a prominent and contentious issue throughout the Scopes Trial. On Friday, July 10, Judge Raulston opened the trial's proceedings by calling on the Rev. Lemuel Cartright, whose prayer admonished listeners "to be loyal to God." Cartright's seven-minute prayer, which John Scopes described as "interminable," prompted Clarence Darrow to turn to Scopes' friend Whitney Morgan and ask, "Does he always do that?" On the following Monday, the Rev. M.H. Moffett opened proceedings with another fundamentalist prayer, this one praising "God our Father, thou who art the creator of the heaven and the earth and the sea and all that is in them." Darrow later noted that he had never during his 40 years as a lawyer "heard God called in to referee a court trial" (Darrow, 1932).

On Tuesday, July 14, before the Rev. A.C. Stribling of Dayton's Cumberland Presbyterian Church could deliver the opening prayer, Darrow objected to opening court-sessions with prayer, telling Judge Raulston, "I do not object to the jury or anyone else praying in secret or in private, but I do object to the turning of the courtroom into a meeting house.... You have no right to do it." Darrow reminded Raulston, "There is no question about the religious character of these proceedings" and objected to having the jury present when his objection about prayer was being discussed. When defense attorney Dudley Field Malone added that the prayers increased "the atmosphere of hostility to our point of view," prosecutor Tom Stewart responded that there is no conflict between science and religion "insofar as the merits [of the case] are concerned" and that the prosecution believed "it is quite proper to open this court with prayer if the Court sees fit to do it ... the agnostic counsel for the defense is foreign to the thoughts and ideas of the people who do not know anything about infidelity and care less." Malone and defense attorney Arthur Garfield Hays answered that they were not agnostic. Stewart then reminded Malone, "This is a God-fearing country," to which Malone noted that it was no more God-fearing than New York, from which he came. Following a heated discussion, Judge Raulston announced that he had "no purpose except to find the truth and do justice" and overruled Darrow's objection because "it has been my custom since I have been judge to have prayers in the courtroom when it was convenient, and I know of no reason why I should not follow up this custom." (At the time, the Tennessee legislature and Supreme Court also opened their proceedings with prayers.) The Reverend Stribling, who considered Darrow "a blatherskite of a pen pusher," then prayed for "a reverence to the Great Creator of the world." Darrow insisted that the court transcript include his objection to the daily prayers, and Raulston agreed. The courtroom debate about prayer consumed much of the morning (Goetz, 1997).

That afternoon, Hays—over prosecutors' objections—read a petition asking Judge Raulston to "select the officiating clergymen [giving prayers] from among other than fundamental churches." Hays then urged the court to hear prayers by "men who think that God has shown his divinity ... in the book of nature, quite as much as in the book of the revealed Word." When Judge Raulston referred the petition to Dayton's association of pastors, several spectators in the courtroom applauded and chuckled, for they knew that most local pastors were fundamentalists. The next day (i.e., July 15), many people were surprised when the Reverend Stribling reported that the local committee of pastors had chosen Unitarian pastor Charles Potter to lead the opening prayer.

Neal, who was upset that Stewart had not apologized for calling Darrow an agnostic, then noted that any prayer "must necessarily be one of one particular faith." After Sue Hicks claimed, "There is no religious controversy in this case," the frustrated Stewart complained that the discussions were a waste of time, adding "I would like to get done with this thing."

On July 16, the Rev. J.A. Allen of Nashville's Lindsley Avenue Church of Christ[35] opened court with a prayer asking that "thy word may be vindicated." The next day, Judge Raulston called on Rabbi Jerome Mark of Knoxville's Beth-El Temple to lead a prayer. Mark was not there, so the prayer was led by the Rev. Charles Eastwood. On Monday, July 20, the opening prayer was delivered by the Reverend Standefer, and the Rev. R.C. Camper of Chattanooga opened court on the trial's final day (July 21). After Scopes was convicted, the Scopes Trial ended with a benediction by the Rev. Charles Jones of Dayton's First United Methodist Church (South). Days later, Jones officiated Bryan's memorial service just a few blocks away.

Sources

Darrow, Clarence. 1932. *The Story of My Life*. New York: Charles Scribner's Sons, 259.
Goetz, David. 1997. The prayer debate: Who could pray at the trial? *Christian History*, 16 (3), 12.

Presley, James Wright
b. 1930

James Wright Presley was born on January 13, 1930, in Nash, Texas. After skipping two grades, he graduated as valedictorian from high school in Redwater, Texas, at age 16. Presley earned a B.A. degree in history from East Texas State Teachers College in 1950, after which he served in the U.S. Air Force from 1950 to 1953. Presley then earned an M.A. from the University of Texas (1955) and did further graduate work at Mexico City College (now the University of the Americas) in Mexican history and anthropology. In 1991, he received a Ph.D. in U.S. history from the University of Texas at Austin.

In 1959, while working as a reporter for the *Shreveport Times*, Presley met John Scopes—then working for United Gas—at a late-night party. Over the years, Presley and Scopes became close friends, and in 1963 Presley wrote a feature about Scopes for *American Weekly*. Presley later recalled, "Around that time it occurred to me that somebody—not necessarily I—ought to write a book about Scopes" (Mahoney, 1966). Soon thereafter, Scopes began working with Presley on his memoirs, which became *Center of the Storm: Memoirs of John T. Scopes* (1967). This book, Presley's first, became a regional bestseller; it eventually went through four printings and was nominated for the Francis Parkman Prize of the Society of American Historians.

Presley has been a full-time writer since 1962. His career has included work at several newspapers, including the *Texarkana Gazette* (1947–1949), *Amarillo Globe Times* (1956), *Shreveport Times* (1958–1959), and *San Antonio Light* (1959–1961, 1969). His work has also garnered two nominations for a Pulitzer Prize, one in 1970 for articles about regional pollution and environmental issues, and another in 1978 for his book *A Saga of Wealth: The Rise of the Texas Oilmen*, which included a cameo appearance by Scopes.

When Scopes died in 1970, Presley was a pallbearer at Scopes' funeral. A late arrival caused by car trouble prevented Presley from taking Scopes' casket into the church, but Presley did help carry Scopes out of the church, a bit of symbolism that, Presley admitted, "John would have liked." At the time of his death, Scopes planned to write another book with Presley, but they never began that work.

In 2022, Presley was a freelance writer and environmentalist. He has written nine books, more than 100 articles, and has won several awards for his work. Presley lives with his wife Fran in Texarkana, Texas.

Sources

Mahoney, Bob. 1966. Ex-newspaperman, Scopes write book. *Shreveport Journal*, February 11, 3.
Scopes, John T., and James Presley. 1967. *Center of the Storm: Memoirs of John T. Scopes*. New York: Holt, Rinehart and Winston.

Press Passes

At the Scopes Trial, reporters issued Criminal Court Room Press Passes were given preferential entrance in the courtroom. The passes were issued by Scopes' prosecutor Gordon McKenzie, the son of prosecutor Ben McKenzie.

McKenzie issued the Scopes Trial's first press pass (i.e., Press Pass No. 1) to *Chattanooga News* reporter Nellie Kenyon, who wrote the first story from Dayton about Scopes' arrest. McKenzie issued Press Pass No. 112 to photographer Henry W. Seavey, who was in Dayton to cover the Scopes Trial for Underwood & Underwood, a pioneering company in news photography. Seavey, who claimed that "we didn't just take pictures; we made them," had been hired by trial instigator F.E. Robinson to document the trial's events (Guffey, 2021). Although many press passes were issued for the Scopes Trial, this one was signed by most of the trial's principals, including the defendant, John Scopes; defense attorneys John R. Neal, Dudley Field Malone, and Clarence Darrow; Attorney General Tom Stew-

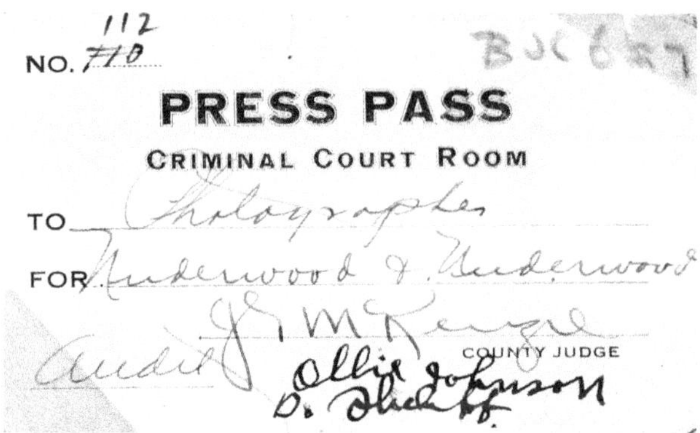

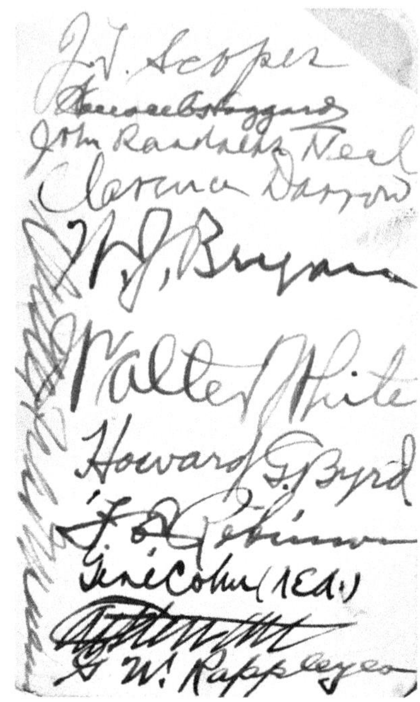

Press passes for the Scopes Trial were issued by Gordon McKenzie, one of Scopes' prosecutors. This pass was issued to photographer Henry Seavey, who had been hired by F.E. Robinson to document the trial's events. Seavey collected the autographs of many of the trial's participants on his pass (RCHGS).

art; prosecuting attorneys Wallace Haggard, Gordon McKenzie, and William Jennings Bryan; Rhea County School Board Chairman F.E. Robinson; reporter Gene Cohn of the Newspaper Enterprise Association (a syndication service for newspaper articles and comics); deputy sheriff Ollie Johnson; Rhea County School Superintendent Walter White (Guffey, 2017); Rev. Howard Byrd; George Rappleyea; and Deputy Sheriff Ollie Johnson. Seavey, who covered the Leopold and Loeb trial and several famous sporting events, remains a source of many famous images of the Scopes Trial (Guffey, 2021).

Source

Guffey, Pat Hawkins. 2007. A badge of history. *Rhea Herald News*, July 12, 1, 2.
______. 2021. History in your hands. *Rhea Herald News*, July 18, 5.

Price, George McCready
1870–1963

George Edward (later McCready) Price was born on August 26, 1870, in Havelock, New Brunswick, Canada. When he was young, Price became a Seventh-Day Adventist and, after graduating from high school at 15, began selling Adventist books. Price attended Battle Creek College (now Andrews University) from 1891 to 1893, and in 1896 enrolled in a teacher training program at what is now the University of New Brunswick. In 1902, while teaching high school in the fishing village of Tracadie, New Brunswick, Price completed his first book, *Outlines of Modern Christianity and Modern Science*,[36] which denounced evolution as being "utterly subversive of civilization and religious liberty." Price later moved to New York to seek work as a writer, but ended up as a construction worker at Adventist church headquarters in Washington, D.C. From 1907 to 1912, Price taught at what is now Loma Linda University, where he received a B.A. In 1920, he was given a M.A. from Pacific Union College. Despite his meager amount of formal education, Price's students liked his courses and were especially impressed with his vocabulary. As one student noted, Price was "the man who swallowed the dictionary and remembered all of it" (Clark, 1966).

In addition to his moral and philosophical objections to evolution, Price advocated "flood geology," which was based on a literal reading of Genesis, including a six-day, 144-consecutive-hours-long Creation Week that occurred "some six or seven thousand years" ago. Price rejected the basic tenets of modern geology, believing instead that the geological record had been produced by a worldwide flood. Price spent his life refuting the "invariable order of fossils" found in the geological record, and claimed in his 726-page *The New Geology* that fossils occur in no predictable order in the geological column (Price, 1923). Price's work was praised by many creationists, but was rejected by mainstream geologists and others.

Price, the most visible young-Earth creationist of his era, became what an editor of *Science* magazine called "the principal scientific authority of the Fundamentalists" (Price, 1926). In his *Illogical Geology: The Weakest Point in the Evolution Theory*, Price even offered $1,000 to anyone who could prove that one kind of fossil is older than another. Price, who claimed that evolution is "a most gigantic hoax," never paid up, and his *The Phantom of Organic Evolution* was a top-seller at T.T. Martin's bookstore during the Scopes Trial. In that book, Price claimed that Satan's "mixing and crossing of the races of mankind" involved degeneration that produced Negroes, Mongolians, and perhaps apes. According to Price, demons, not natural selection, had created many species (Numbers, 2006).

Fundamentalists such as John Roach Straton and Frank Norris praised Price's *The New Geology* as a "great monumental" book that "utterly demolishes" evolution (Numbers, 2006). Bryan believed that Price was an "outstanding scientist" and invited him to Dayton to help prosecute Scopes. However, Price—then teaching at Stanborough Missionary College just outside of London—did not come. Instead, Price told a *London Times* reporter during the trial that Americans knew more about evolution than did their British counterparts and were therefore "much better prepared to say that the evolution idea is at best only a theory" (Price, 1925).

Price urged Bryan to avoid discussing science at the trial (Numbers, 2006). When Bryan mentioned Price in his testimony, Darrow denounced Price as "the only human being in the world so far as you know who signs as a geologist that believes like you do ... every scientist in this country knows [Price] is a mountebank and a pretender and not a geologist at all." After the trial, Price turned against "poor Bryan, with his day-age theory of Genesis, [who] was so piteously overwhelmed and defeated by the battery of evolutionary scientists under the clever leadership of the atheist Clarence Darrow ... [Bryan] really didn't know a thing about the scientific aspects of the case." In 1942, Price claimed that the Scopes Trial was "a turning-point in the intellectual and religious history of mankind," a defeat for fundamentalism, and had "undoubtedly [been] planned long ahead by the devil ... the general results were a triumph for the devil's cause such as the world has seldom seen before or since" (Price, 1942; Alston, 2003).

Throughout his life, Price condemned theistic reconciliations of Genesis with an ancient Earth (i.e., such as those held by most of fundamentalism's leaders in the early 1920s) as a "truly pagan ... heathen religion" and "the devil's counterfeit" (Price, 1954). Price demanded that "the geologist and biologist hold steady to the exact wording [of the Bible] without dodging or quibbling or without any pleas for wrong translation or interpolations or imperfections of the record" (Clark, 1966). Price believed that the last reform in this world would be "a return to the simple truth of creation," which Price believed involved processes no longer in operation.

In 1935, Price co-founded the short-lived Religion and Science Association, which repudiated an ancient Earth. That was followed in 1938 by the Deluge Geology Society, whose members had to endorse a six-day creation and a worldwide flood that produced "the major geological changes since creation." That society, after searching for Noah's Ark, disappeared in the late 1940s. In 1938, after teaching for five years at Walla Walla College, Price moved to Loma Linda, California, where he spent the rest of his life.

Price wrote more than 20 books and 800 articles about creationism, many for laypeople. Although others labeled Price's *The New Geology* "a classic of pseudoscience," Price believed his book's ideas were "far superior to any hitherto offered ... [and] the only one worthy of the time and attention of fact-loving men and women" (Clark, 1966). In the 1950s, Martin Gardner noted, "Among 20th century Protestant opponents of evolution, one man and one alone stands head and shoulders above all others. He is the 'geologist' whom Bryan cited as his chief authority at the famous Scopes Trial in Tennessee.... Almost every fundamentalist attack on evolution in the last three decades has drawn its major ammunition from his writings. He is, in fact, the last and greatest of the antievolutionists" (Gardner, 1952). Near the same time, Baptist theologian Bernard Ramm—who dismissed Bryan's "miseries at the Scopes Trial" as part of a "sordid history"—praised Price as "fundamentalism's leading apologist in the domain of geology.... The influence of Price is staggering" (Ramm, 1954).

Price's ideas became a foundation (without appropriate citation) of *The Genesis Flood*, the founding document of modern creationism. Henry Morris—a coauthor of *The Genesis Flood* who earned a Ph.D. from the University of Minnesota in hydraulic engineering to better defend the worldwide flood described in Genesis 7–8—described reading Price's *The New Geology* as "a life-changing experience." One of Price's students at Andrews University, Frank Lewis Marsh, became the first Seventh-Day Adventist to earn a Ph.D. in biology. Price's grandson, actor John Shelton, named one of his children Darwin to "balance everything out."

Following a six-week illness, Price died in Loma Linda on January 24, 1963, and was buried near there in the Montecito Memorial Park and Mortuary Cemetery. In 1973, Andrews University named its new biology building after Price, despite the fact that Price had never taught there.

Sources

Alston, Jon P. 2003. *The Scientific Case Against Scientific Creationism*. New York: iUniverse, 16.

Clark, Harold W. 1966. *Crusader for Creation: The Life and Writings of George McCready Price*. Mountain View, CA: Pacific Press, 32, 48–49, 54.

Gardner, Martin. 1952, 1957. *Fads and Fallacies in the Name of Science*. New York: Dover, 9, 127, 132.
Numbers, Ronald L. 2006. *The Creationists: From Scientific Creationism to Intelligent Design*. Cambridge: Harvard University Press, xiii, 100–102, 115–116.
Price, George McCready. 1923. *The New Geology*. Mountain View, CA: Pacific Press, 296, 637–638.
_______. 1925. The evolution trial. *London Times*, July 16, 10.
_______. 1926. Letter to the editor of *Science* from the principal scientific authority of the Fundamentalists. *Science*, 63 (1627), 259.
_______. 1942. What Christians believe about Creation. *Bulletin of Deluge Geology*, 2, 76.
_______. 1954. *The Story of Fossils*. Mountain View, CA: Pacific Press, 39.
Ramm, Bernard. 1954. *The Christian View of Science and Scripture*. Grand Rapids: Wm. B. Eerdmans, 180.

Race and Racism

The Scopes Trial did not invoke race explicitly, but race was an important part of the fundamentalists' campaign that produced the trial. Darwin did not believe that human races are discrete species, but countless scientists used his ideas to promote White supremacy as biological fact. Antievolutionists warned that evolution was ruining Southern morals, destroying "Southern Womanhood," and threatening White supremacy by making "a Negro as good as a white man" (Cash, 1941). Evolution implies kinship and common origins, and Whites accepting that would "have to admit that there is no fundamental difference between themselves and the race they pretend to despise" (If monkeys could speak, 1925; Moran, 2003). That is, if Blacks and Whites shared a common ancestor, as evolution suggests, then White supremacy would be arbitrary instead of sanctioned by God. Fundamentalist leaders such as Frank Norris made this explicit when he reminded his followers, "Well, God didn't make them [i.e., Blacks] that way [i.e., equal]. It's hard to go against God's laws," and John Roach Straton believed that attempts to "improve" African Americans were doomed to fail because of their inherent "tendency to immorality and crime" (Landslide in Texas election, 1952; Straton, 1900). Norris and many other fundamentalists promised that the country's morality could only be saved by Biblical inerrancy, White supremacy, and opposition to ideas such as evolution that seemed to undermine the Bible. However, when a fundamentalist at the Scopes Trial tried to sell a pamphlet claiming that Blacks are not human, *Time* magazine reported, "The Tennessee mountaineer is no 'nigger hater.' In fact, he fought in the Union army to help free the Negro.... He shows no desire to listen to attacks on the Negroes" (Brown, 1997).

The links of science and racism did not begin with Darwinism, but Darwin's ideas popularized them. Prominent scientists such as Henry Fairfield Osborn and Edwin Grant Conklin—both of whom advised Scopes' defenders before and during the trial—conflated evolution and eugenics to produce racial hierarchies. Conklin invoked evolution, biological determinism, and the "best races" to promote social inequalities, while Osborn believed that the genus *Homo* included several species (e.g., "Caucasian, the Mongolian, and the Negroid") that were separated by "spiritual, intellectual, moral, and physical characters.... The standard of intelligence of the average adult Negro is similar to that of the 11-year-old youth of the species *Homo sapiens*" (Osborn, 1926). Eugenics would preserve this racial sorting by intervening in reproduction and, in the process, plan the "lower races" out of existence.

Scientists' alleged evidence for "scientific racism" was used to determine how people viewed, studied, and exploited differences among people. Racial hierarchies were common in science and science textbooks; for example, George Hunter's *A Civic*

Biology—the textbook that John Scopes used in Dayton—claimed that natural selection had produced "five races or varieties of man" that culminated with "the highest type of all, the Caucasians."

In the 1920s, campaigns to ban the teaching of evolution often had strong racial overtones. For example, influential preacher and Klansman Benjamin Bogard's (1868–1951) American Anti-Evolution Association welcomed everyone except "Negros [sic], and persons of African decent [sic], Atheists, Infidels, Agnostics, such persons as hold to the theory of Evolution, habitual drunkards, gamblers, profane swearers, despoilers of the domestic life of other desecrators of the Lord's Day, and those who would depreciate feminine virtue by vulgarly discussing sex relationship" (de Camp, 1968). In 1928, as a result of Bogard's campaign, Arkansas became the first and only state to outlaw the teaching of evolution in its public schools with a public referendum. When Susan Epperson later successfully challenged the Arkansas law, she received letters condemning her and civil rights laws, and promising that "if I was a teacher, the first nigger that walked into my classroom I would walk out." Another letter, dated "Easter Sunday 1966," closed with "P.S.—I'm white, too" (Moore and Chung, 2005).

African Americans were interested in the Scopes Trial (Moran, 2003, 2012; Robinson, 2018). Intellectuals such as W.E.B. Du Bois (1868–1963), many of whom viewed Scopes as a victim of Southern oppression, accepted evolution and opposed the Butler Act because they linked education and intellectual freedom with social progress and their fight against White supremacy; as DuBois noted, "Dayton, Tennessee, is America: A great, ignorant simple-minded land, curiously compounded by brutality, bigotry, religious faith, and demagoguery" (DuBois, 1925). Black educator and journalist William Pickens (1881–1954) dismissed the Butler Act by noting that "the law of evolution cannot be overthrown by the wisdom of a state where they burn men alive" (Pickens, 1925). Just weeks after the Scopes Trial, the popular African American weekly *Chicago Defender* (est. 1906) urged its readers to "make [their] best showing by learning the meanings of unfamiliar words." The first word on its list was "evolution" (Vocabulary terms, 1925).

The Rev. W.H. Moses, who attended the Scopes Trial with other African American preachers, hoped that the trial would show "that Christianity is strengthened by science" (Moran, 2013). However, some other Black pastors—who pledged allegiance to the Bible—were hostile to evolution because of how it was used by scientists and others to promote racism. For example, just two months after the Scopes Trial, the National Baptist Convention—the largest African American religious organization in the United States—passed resolutions denouncing evolution and the KKK (Moran, 2012). Although many of the monkey-based cartoons in newspapers ridiculed fundamentalists, some had a more serious tone. For example, a cartoon in *Chicago Defender* showed a pair of horrified monkeys clinging to each other in disbelief as they watched a lynching below (Moran, 2002, 2003).

Many African American Christians were fundamentalists who overlooked Bryan's racist views because of his strong stand for the Bible. For example, Chattanooga's the Rev. George Washington Sandefur announced, "The Negro commends.... Bryan for his solid stand for the fundamentals of our faith," adding that he "did not know a single minister among my race in Tennessee who is an evolutionist.... We have followed the white race for good and evil, from Africa on through the centuries to the present, but if they are going after the modern evolution business, here is where we part" (Callaway, 1925). Few Black journalists supported Bryan, for he—like many Southern sympathizers—supported Jim Crow and claimed to be "a member of the greatest of all the races—the Caucasian race" (Bryan and Bryan, 1925; Robinson, 2018). Bryan excused his beliefs by noting that "the white people of the South, if in control, would be more apt to deal justly with the blacks than the blacks would be, if in control, to deal justly with the whites." Although Bryan claimed that "blacks were entitled to equal protection of their rights," he added that Blacks "could not formulate laws and administer government as well as the whites." Blacks had improved their lives because they "have been associated with and helped by white people. Slavery among the whites was an improvement over independence in Africa. The very progress that blacks have made, when—and only when—brought into contact with the whites ought to be a sufficient argument in support of white supremacy—it ought to be sufficient to convince even the blacks themselves.... Anyone who will look at the subject without prejudice will know that white supremacy promotes the highest welfare of both races" (Bryan, 1923). Bryan was condemned by several Black editors because "he taught by both word and action that infamous doctrine of the devil, White supremacy. The leaders bent on lynching Negroes and instigators of discrimination laws in the South were invariably men he was pleased to call his followers" (William Jennings Bryan, 1925).

As early as 1901, Darrow had advocated interracial marriage, condemned racial prejudice, and noted "how the colored race was particularly subjected to that injustice and oppression" (Darrow, 1901). However, just four months after the Scopes Trial, he described slavery as a necessary evil because it had introduced Blacks to civilization. His comments resembled those made earlier by Bryan: "Without slavery, your race would never have had its chance for civilization.... I think that civilization is worth the price we have to pay for it, what the white man paid in his upward struggle" (Darrow says, 1925). The next year, Darrow launched a racist and derogatory attack on European immigrants. After complaining that "undesirables" were reproducing "at an indecent rate," he lamented, "Here in our own country, which was settled by the Nordics after the Indians were driven out, the superior race is fast going down before the misfits of inferior races" (Darrow, 1926).

Sources

Brown, Rodger Lyle. 1997. *Ghost Dancing on the Cracker Circuit: The Culture of Festivals in the American South*. Jackson: University of Mississippi Press, 136.

Bryan, William Jennings. 1923. Bryan says North would act as South on Negro question. *New York Times*, March 18, Special Section, XXI.

______, and Mary Baird Bryan. 1925. *The Memoirs of William Jennings Bryan*. Philadelphia: John C. Winston, 330.

Callaway, T.W. 1925. Monkey is kin only in court. *Chattanooga Times*, July 17, 3.

Cash, W.J. 1941. *The Mind of the South*. New York: Alfred A. Knopf, 339.

Darrow, Clarence. 1901. The problem of the Negro. *International Socialist Review*, 2 (5), 29, 324.

______. 1926. The eugenics cult. *American Mercury*, 8 (30), 129.

Darrow says dry act hasn't killed "appetite." 1925. *Detroit Free Press*, November 9, 1.

de Camp, L. Sprague. 1968. *The Great Monkey Trial*. Garden City, NY: Doubleday, 475.

DuBois, W.E.B. 1925. Scopes. *Crisis*, September, 218.

If monkeys could speak. 1925. *Chicago Defender*, May 23, 12.

Landslide in Texas election for Dwight D. Eisenhower. 1952. *Fundamentalist*, August 1, 2.

Moore, Randy and Carl Chung. 2005. "P.S.—I'm white too": The legacy of evolution, creationism, and racism in the United States. *Science Education Review*, 4 (2), 50.1–50.14.

Moran, Jeffrey. 2002. *The Scopes Trial: A Brief History with Documents*. Boston: Bedford/St. Martins.

______. 2003. Reading race into the Scopes Trial: African-American elites, science, and fundamentalism. *Journal of American History*, 90, 891–911.

———. 2012. *American Genesis: The Evolution Controversies from Scopes to Creation Science*. Oxford: Oxford University Press, 73.
Osborn, Henry Fairfield. 1926. The evolution of human races. *Natural History*, 89 (April 1980), 129.
Pickens, William. 1925. Monkey business. *New York Amsterdam News*, July 22, 16.
Robinson, Shantá R. 2018. A crusader and an advocate: The Black press, the Scopes Trial, and educational progress. *Journal of Negro Education* 87 (1), 5–21.
Straton, John Roach. 1900. Will education solve the race problem? *North American Review*, 170 (523), 785–802.
Vocabulary terms. 1925. *Chicago Defender*, August 29, A8.
William Jennings Bryan. 1925. *New York Amsterdam News*, July 29, A1.

Rappleyea, George Washington 1894–1966

George Washington Rappleyea (often misspelled "Rappelyea") was born on July 4, 1894, in New York City to George M. and Marian Rogers Rappleyea. As a child, Rappleyea took art lessons and sold newspapers at Times Square. When he was 18, Rappleyea earned a degree in civil engineering from Ohio Northern College in Ada, Ohio. On May 26, 1917, Rappleyea was called into service for World War I at Ft. Oglethorpe, Georgia, where he worked in the Engineering Corps. While in the military, Rappleyea applied what he learned in art classes to produce military sketches. Three months after enlisting, Rappleyea was honorably discharged at the "Convenience of Govt" without having served overseas.[37] Rappleyea, who later claimed to have multiple doctorate degrees, then began working as a geologist and engineer, but he continued to publish articles about military sketches (e.g., Rappleyea, 1921, 1943).

While surveying a Tennessee farm in 1922, Rappleyea was bitten by a snake and hospitalized, at which time he met Ova "Precious" Corvin, his nurse and future wife. Ova's brother, who owned a grocery store in Dayton, helped Rappleyea get a job managing the Cumberland Coal & Iron Company, which was liquidating the remnants of the once-thriving DCIC. The failure of DCIC in 1913 had staggered Dayton's economy, and Rappleyea and others were looking for ways to improve it.

On May 4, 1925, the energetic Rappleyea—a member of the ACLU—noticed an ad placed by the ACLU in the *Chattanooga Times* searching for a Tennessee teacher to test the newly passed Butler Act. Rappleyea took the ad to Robinson's Drug Store and, after conferring with several people who regularly met there, convinced coach and science/math teacher John Scopes to test the law. Rappleyea—who believed that the Butler Act was "un-American" and "a religious law"—swore out a warrant for Scopes' arrest on May 5, and charges were filed the next day. According to F.R. Rogers, who was then a pharmacist at Robinson's, Rappleyea "didn't even know Scopes.... Rapp called him 'Scoops' instead of Scopes" (In Scopes Trial, figures recall old memories, 1960).

Rappleyea and his wife lived on a 40-acre farm just outside of town and attended the nearby Five Points Methodist Church, where he was baptized and made superintendent of the Sunday School. Rappleyea accepted evolution, and his pastor, Howard Byrd, convinced him that evolution was not a threat to Christianity. Rappleyea—described variously as "an untidy little person with rather ill-tended teeth" and "a combative little rooster with ... a habit of getting into arguments" (Stone, 1943; Setting the stage, 1925)—was allegedly infuriated when, at the funeral of a young boy, a fundamentalist preacher claimed that the boy was suffering in hell because he had not been baptized.

George Rappleyea, shown here in a laboratory of Cumberland Coal & Iron Company, instigated the Scopes Trial (Freddie Brewer).

Rappleyea told reporters that his religion involved "doing what good I can to help the city, state, country, and humanity along.... I don't believe in evangelical Christianity; before the preacher, crying on your knees today, forgotten tomorrow." Rappleyea was happy when he learned that Bryan would help prosecute Scopes, telling a reporter, "The Lord has delivered him into my hands" (Seek services of H.G. Wells, 1925). Rappleyea then showed reporters a letter from his mother, telling her son, "I do not believe in evolution, and if I did I would not have the face to tell anybody that I did. You always had lots of book sense, but never any common sense." When questioned about the letter, Rappleyea asked, "That can I do about it? I can't give up my convictions" (Was converted through science, 1925). Rappleyea, like many others in Dayton, hoped that Scopes' trial would boost the local economy.

After Scopes' arrest, several players on Scopes' football team began planning to "beat the tar out of" Rappleyea for swearing out the warrant against Scopes, but Scopes—"a hero of the students"—convinced them to cancel their plans (Anderson, 1925). Rappleyea promoted the upcoming trial; at one point, he even staged a fake fist-fight with local barber Thurlow Reed for reporters. Rappleyea and his wife renovated the abandoned, 18-room "Mansion" to house some of Scopes' attorneys and the scientists coming to Dayton to testify as expert witnesses. Throughout the trial, Rappleyea—who often talked with reporters and led tours of Dayton—was "here, there, and everywhere ... a whole entertainment committee in one man" (Curtis, 1965). Rappleyea was Scopes' original complainant (i.e., prosecutor), but on May 25 he was replaced by Rhea County

School Superintendent Walter White, who was more sympathetic to the fundamentalists' cause. Rappleyea often sat with the defense when court was in session; although had earlier admitted to reporters that "Dr. Neal accepted Darrow's help in the case," he added, "I wish he had not" (Dayton jolly, 1925).

Soon after the trial, Rappleyea—lamenting that "you cannot imagine the inglorious position of being branded as an evolutionist in the town of Dayton"—left town (Rappleyea tells inside story, 1926). He joined the Adams Lecture Bureau to speak about "The Battle of the Century" and "When Truth Is a Crime" for $100 per speech, but there was relatively little interest in his talks. His plans to open a liberal college in Dayton also never materialized (Rappleyea tells inside story, 1926; de Camp, 1968). Before leaving Dayton, Rappleyea—who attended Scopes' appeal before the Tennessee Supreme Court—donated Cumberland Coal and Iron Company's remaining laboratory equipment and a ton of coal to William Jennings Bryan University.

In 1939, Rappleyea was working in New Orleans, Louisiana, as a vice-president and assistant to boatbuilder Andrew Jackson Higgins, the president of Higgins Industries (est. 1930). By 1943, Higgins's company had produced 12,964 (i.e., 92 percent) of the Navy's 14,072 boats, and Rappleyea had helped train more than 30,000 people to operate and maintain Higgins boats, including those used by Allied troops in World War II to storm beaches at Normandy, Iwo Jima, Okinawa, and elsewhere (Rappleyea, 1943).[38] While with Higgins, Rappleyea was described as being "famous in his own right" and "known far and wide" in yachting circles. In 1943, Rappleyea—apparently forgetting his years as a mining engineer in Dayton—claimed that he had "devoted a life-time to the teaching of the subject of navigation" (Rappleyea, 1943).

In 1946, after retiring at Higgins, Rappleyea was working as treasurer of Marsalis Construction, a company in New Orleans founded by retired military pilot William Marsalis. Although Rappleyea claimed that the company was in "the mahogany business in British Honduras," Marsalis confessed that the company's only purpose was to support a coup to make Cuba the 49th state. Marsalis Construction had procured tanks, bomb-cases, high-explosives, machine guns, pistols, shotguns, thousands of rounds of ammunition, landing ships, P-38 fighter planes, and teargas grenades, all of which were stored in a warehouses and vacant homes in Gulfport, Mississippi. According to Marsalis, Rappleyea—who had allegedly recruited pilots to bomb Havana—"was going to have me shot in Cuba 'accidentally' after the coup. So I fixed it up to dispose of Rappleyea when he reached Cuba" (Huie, 1964). Law enforcement preempted either of the killings, and on March 2, 1947, Rappleyea and others were arrested in New Orleans for violating the National Firearms Act. After posting $5,000 bond, Rappleyea claimed that the planes—all designed and intended for aerial combat—were for "aerial mapping of rivers in Central America," and the machine guns and explosives for "hunting in South America" (Huie, 1964).

On March 31 of the next year, just after his trial started, Rappleyea pled guilty in federal court in Biloxi, Mississippi, to conspiring to ship arms and ammunition to British Honduras without a license from the State Department. After announcing, "I am probably giving you … a lighter sentence than most other judges would give," Judge Sidney Mize sentenced Rappleyea to one-year-and-one-day in the federal prison in Texarkana, Texas. Rappleyea also received a five-year suspended sentence for possessing unregistered firearms (Three sentenced, 1948). Although John Scopes lived only 75 miles away in Shreveport, he never visited Rappleyea in prison, and he later instructed biographer James Presley not to mention Rappleyea's incarceration in his memoir.

After being released from prison, Rappleyea returned to Southport, North Carolina, where he had earlier owned an electric company and an ice house. Rappleyea directed the Tropical Agricultural Research Laboratory, from which he registered several patents, including one for "Plasmofalt," an inexpensive building material named for its plastic, molasses, and asphalt ingredients. Rappleyea's invention, which was featured in *Popular Mechanics*, was marketed as a way of quickly making inexpensive houses, driveways, and airstrips for planes on sandy islands; a five-room house atop a molasses slab would cost only $1,000 (plus wiring and plumbing; Day, 1951; Whitmire, 1951). The U.S. Marine Corps subsidized some of the initial production and tests conducted by Rappleyea in his backyard; parts of Rappleyea's house and garage were made of Plasmofalt (The way it was, 1976). Rappleyea planned to open a Plasmofalt factory in New Jersey, but in 1955 he gave his patent to Ohio Northern College. The patent was never exploited. In 1962, Rappleyea was still directing the Tropical Agricultural Research Laboratory, which had moved to Miami, Florida.

When Rappleyea—a member of the First Humanist Society of New York—died at age 72 in Miami on August 29, 1966, his death certificate listed his occupation as "Boat Builder." He is buried in Arlington National Cemetery. His wife, Ova ("Precious," 1899–1979), who was prominent during the Scopes Trial, died on December 6, 1979, and is buried in Spivey Cemetery near Dayton.

Sources

Anderson, Paul Y. 1925. The drugstore argument that swelled into a national debate. *St. Louis Post-Dispatch*, May 31, 17.

Curtis, W.C. 1965. The evolution controversy. In Tompkins, Jerry R. 1965. *D-Days at Dayton: Reflections on the Scopes Trial*. Baton Rouge: Louisiana State University Press, 74–85.

Day, Michael. 1951. You'll build your next house of molasses. *Popular Mechanics*, 96 (3) (September), 110–114, 248, 250, 252.

Dayton jolly as evolution trial looms. 1925. *Chattanooga Daily Times*, May 21, 1.

de Camp, L. Sprague. 1968. *The Great Monkey Trial*. Garden City, NY: Doubleday.

Huie, William Bradford. 1964. *The Hiroshima Plot*. New York: G.P. Putnam's Sons, 71–79.

In Scopes Trial, old figures recall old memories. 1960. *Knoxville News-Sentinel*, July 17, 1960, B-10.

Rappleyea, George W. 1921. Military sketching as a hobby. *The Military Engineer*, 13 (70), 362.

______. 1943. *Navigation Wrinkles for Combat Motor Boats*. New Orleans: Higgins Industries, n.p., preface.

______. 1944. On the Higgins landing boats. *The Eureka News Bulletin*, 3, 84–92.

Rappleyea tells inside story of Scopes Trial, scouts accepted version. 1926. *Mobile Sunday Register*, January 24, 1, 3.

Seek services of H.G. Wells in aid Scopes. *Chattanooga Daily Times*, May 15, 11.

Setting the stage for a big battle between science and religion. 1925. *Kansas City Star*, June 7, 41.

Stone, Irving. 1943. *Clarence Darrow for the Defense*. Garden City, NY: Doubleday, 434.

Three sentenced on arms charge in Biloxi court. 1948. *Sun Herald (Biloxi, MS)*, March 30, 1.

Was converted through science. 1925. *Chattanooga Times*, May 21, 2.

The way it was. 1976. *The State Post Pilot* (Southport, NC), September 29, 22.

Whitmire, Paul. 1951. Tar for the heels of the Marines. *The State Weekly Survey of North Carolina*, 19 (30), 3, 19.

Raulston, John Tate
1868–1956

John Tate Raulston was born on September 22, 1868, in Fiery Gizzard Cove in Marion County, Tennessee. After attending country schools and U.S. Grant Memorial University (today's Tennessee Wesleyan University), he was admitted to the bar in 1896. He taught school for four years and then began practicing law in Whitwell, Tennessee, just northwest of Chattanooga. Raulston, a Republican, served in the Tennessee state legislature from 1902 to 1904, ran unsuccessfully for the U.S. House of Representatives in 1908, and was elected judge of Tennessee's seven-county Eighteenth Circuit in 1918. Raulston was described as "a popular local attorney of no special competence" (Linder, 2020).

Raulston's most famous case was the Scopes Trial, for which he suggested that Dayton build a stadium that would hold 20,000 "at the very least" spectators. Raulston scheduled the trial to start on July 10, "when all universities and schools will be through their terms of school" and "scientists, theologians, and other school men will be able to act as expert witnesses" (Trial July 10 suits Bryan, 1925). Ironically, during the trial, Raulston banned expert witnesses from testifying before the jury.

Raulston, a Methodist lay minister who had recently preached at a revival in Dayton, opened proceedings each day with prayer. Throughout the trial, the publicity-conscious Raulston granted interviews, posed for pictures, visited with people in the courtroom and at Hotel Aqua (where he stayed during the trial with his family), and often delayed proceedings to accommodate photographers. Raulston quoted scripture during the trial and often deferred to Bryan. After scientist Maynard Metcalf testified on July 15, Raulston barred live testimony from the other scientists brought to Dayton by the defense. Raulston could not conceive how the Butler Act violated teachers' rights because "the relations between the teacher and his employer are purely contractual, and if his conscience constrains him to teach the evolutionary theory, he can find opportunities elsewhere in other schools." On the trial's last day, Raulston ruled that Bryan's testimony in response to Darrow's questions was irrelevant and inadmissible. After Bryan died, Raulston wept and professed his belief in "the God of Bryan." Raulston's wife, Eva, told reporters that Darrow "made light of the [trial], while Mr. Bryan took it seriously." She also complained that she had to stay near her husband throughout the trial because "people wanted to take a picture of him with a monkey on his shoulder ... [I had to] knock the monkeys off him" (Anderson, 1955).

Judge John Tate Raulston was a lay preacher and judge in Tennessee's Eighteenth Circuit who presided at the Scopes Trial (New York Public Library Image No. 97364).

In 1926, Raulston lost his bid for reelection to Democrat Leslie Darr, an attorney from Jasper, Tennessee. Raulston then practiced law with the Raulston, Raulston, & Swafford law firm in South Pittsburg, Tennessee, and preached about fundamentalism. Raulston continued to praise Bryan and claim that evolution "would destroy God's word." In 1934, Raulston sought the circuit judgeship again, but was defeated again by Darr. Three years later, Raulston demanded that children "be protected from the teaching of such insidious doctrines [such as evolution] on the ground that they ... [produce] mental stupidity and spiritual degeneracy" because evolution is "irrational and unworthy of belief." Raulston admitted, "The trial stimulated talk the world over, but as to good [it accomplished]—well, I just don't know" (Harris, 1937).

When he retired in 1950, Raulston urged Tennessee to enforce the Butler Act. However, just before his death, Raulston acknowledged that government should not restrict the teaching of science when doing so doesn't damage students' morals, adding, "It is silly to think all of this universe came by chance and not by design. I had and have a profound belief in the creation of man." Raulston described Bryan as "the most outstanding man at the trial," but claimed that although "Darrow may have been a greater lawyer—from point of reputation," he was "a very pathetic man" because he had "no hope for future life" (Harris, 1937).

In 1950, Raulston—who admitted that the Scopes Trial was "a difficult undertaking for a rural judge"—had a nervous breakdown, followed by a fall that broke his hip. Thereafter, he was confined to his home and grew increasingly frail. The 87-year-old Raulston died on July 11, 1956, at which time John Scopes eulogized him as "a grand old man" who "was quite fair in his handling of the case. He was trying to do what he was elected to do—follow the laws of Tennessee—and trying to do what he thought was right" (Raulston, Scopes Trial judge, 1956). Later, Scopes' witness Winterton Curtis admitted that Raulston had not been "so partial as we thought" during the trial (Curtis, 1965).

Raulston is buried in Cumberland View Cemetery in Kimball, Tennessee.

Sources

Anderson, William B. 1955. Presiding judge recalls moments of historic Scopes "Monkey Trial." *Fort Worth Star-Telegram*, July 4, 14.
Curtis, W.C. 1965. The evolution controversy. Pp. 74–85 in Tompkins, Jerry R. 1965. *D-Days at Dayton: Reflections on the Scopes Trial*. Baton Rouge: Louisiana State University Press.
Harris, Dick. 1937. Judge of Dayton evolution trial is doubtful of benefits derived. *Chattanooga Times*, April 26, 7.
Linder, Douglas O. 2020. Judge John T. Raulston. Accessed September 28, 2020, at https://famous-trials.com/scopesmonkey/2099-raulston.
Raulston, Scopes Trial judge, dies. 1956. *The Knoxville Journal*, July 12, 1, 12.
Trial July 10 suits Bryan. 1925. *Nashville Banner*, May 26, 1.

Reece, Raleigh Valentine
1901–1946

Raleigh Valentine "Hunkey" Reece was born on September 8, 1901. After attending high school in Clearwater, Florida, Reece played football and graduated from Carson-Newman College, which was allegedly the only institution in Tennessee that strictly followed the Butler Act. Reece was then hired by Dayton's Rhea Central High School in 1923 to teach history and coach sports. (The previous year, the football team there had ended its season with no wins and no coach.) Following a losing season, Reece resigned in May 1924 to become a reporter for the *Nashville Tennessean* and was replaced by John Scopes. In late June 1925, as the Scopes Trial approached, Reece wrote Sue Hicks asking for "inside dope ... [and] what the plan of the prosecution is, and anything else pertaining to the trial or any of the principals.... If I can get a good story on this.... I might be sent to Dayton to cover the trial" (Of monkeys and men, 2021).

In August 1925, Reece announced that he was a fundamentalist who believed, "The Virgin birth ... is an unqualified fact, as is the existence of heaven and hell" (Tennessean staff man, 1925). The next day, he was rehired by Rhea Central High School to coach and teach history. At least one student at Rhea Central was unimpressed, complaining to a reporter that "our football team is busted.... What does a fundamentalist know about football?" (Moore, 2022).

When classes began in Fall 1925, Reece missed the first few days of class. In his place was substitute teacher Doris Darwin, a medical student at the University of Louisville. Darwin accepted evolution, but promised, "I won't teach it here. I am no fundamentalist" (Darwin in Scopes' place, 1925). Meanwhile, Tennessee, newspapers referred to the Rhea Central football team as "The Fundamentalists." In the midst of that season, Reece—who was assisted by Scopes' friend McKinley "Mack" Jones—told reporters, "The question of evolution is never referred to any more in Dayton—either in school or by the people" (Raleigh V. Reece, 1925). Early the following year, after another losing season, Reece resigned, allegedly at the request of the school board, following "school entertainment which the [board] did not approve" (Successor to Scopes resigns, 1926).

In 1929, after several years as a newspaper reporter and telegraph operator, Reece began working as a textile manager for American Bemberg Corporation in Elizabethton, Tennessee. In 1942, he joined the military and served in the Army Air Corps.

Reece died on July 19, 1946, at age 43 in the Veterans Administration Hospital in Mountain Home, Tennessee. He is buried in Monte Vista Memorial Park in Johnson City, Tennessee.

Sources

Darwin in Scopes' place for a while. 1925. *Chattanooga Daily Times*, May 15, 11.
Moore, Randy. 2021. Coach Scopes. *NCSE Reports*, 42 (1), 13.
Of Monkeys and Men: Public and Private Views from the Scopes Trial. 2021. University Libraries, University of Tennessee, Knoxville.
Raleigh V. Reece, Scopes' successor, is visitor here. 1925. *Johnson City Chronicle*, October 20, 4.
Successor to Scopes resigns. 1926. *Nashville Banner*, February 1, 4.
Tennessean staff man will succeed Scopes at Dayton. 1925. *Nashville Tennessean*, August 16, 1, 8.

Rhea Central High School
Est. 1906

Rhea Central High School (RCHS) opened in 1906 as Dayton's first public high school. In the summer of 1924, RCHS hired John Scopes to coach and teach science and math for $150 per month. Of these jobs, Scopes considered coaching to be most important (Hardaway, 1960). Scopes was a popular teacher; his football team posted a 4–4–2 record, and his other teams won more games than they lost.

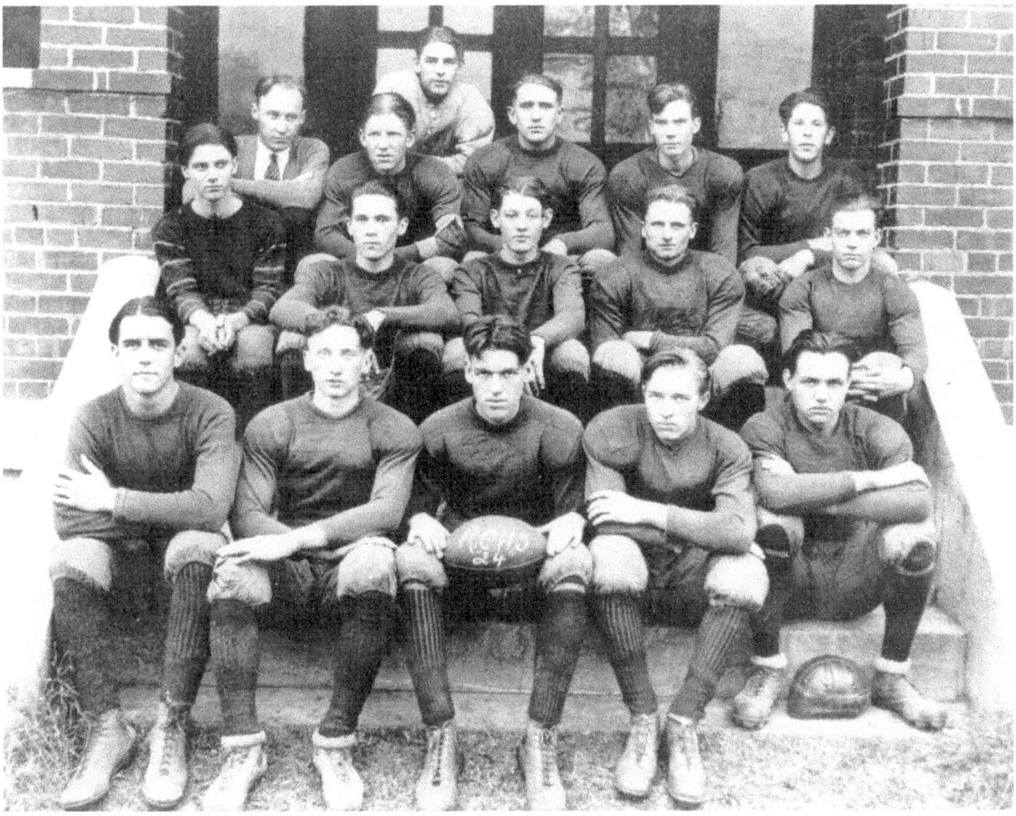

In fall 1924, John Scopes (wearing tie in top row, upper left) coached the Rhea Central High School "Yellow Jackets" football team to a 4–4–2 season. Scopes, who also taught science and math, considered coaching to be the most important part of his job. This photograph shows Scopes posed with his football team on the steps of Rhea Central High School in the fall of 1924 (RCHGS).

When RCHS principal and biology teacher William Ferguson got sick in April 1925, Scopes was the substitute teacher in his classes. It was during this time that Scopes allegedly taught evolution and, in doing so, violated the Butler Act. When he was arrested, Scopes assumed that he would lose his job, but admitted that it "was nothing to lose." Scopes believed that his trial "would be just a little trial like any other lawsuit at the courthouse," but soon lamented that "it looks like a fine mess I got myself into" (Scopes and Presley, 1967).

Scopes' contract ended on May 1, 1925, but he stayed in town to look after two of his injured students and pursue "a beautiful blonde I had somehow previously overlooked" (Scopes and Presley, 1967). During Scopes' trial, Dayton officials considered using RCHS as an emergency hospital, but those plans never materialized. Chicago's WGN, which broadcast the trial, installed four speakers on the courthouse lawn and at each of four other places—RCHS, a local church, a grammar school, and the American Legion Hall—so people could hear the trial.

After the trial, school superintendent Walter White offered to renew Scopes' contract, provided Scopes would "adhere to the spirit of the evolution law." Scopes declined the offer and instead used $2,250 raised by the defense team's scientists to attend graduate school at the University of Chicago. Scopes was replaced at RCHS by Raleigh Reece, a reporter from the *Nashville Tennessean* who rejected evolution.

RCHS was abandoned in 1928 when the school moved into a new building nearby. Two years later, the first classes of William Jennings Bryan University met in the abandoned building in which Scopes had allegedly taught evolution. Bryan University moved to its current location in 1935, after which the old RCHS building was destroyed. Today, a historical plaque stands near the school's original site.

Sources

Hardaway, Sybil. 1960. Dayton key important trophy to Mrs. Scopes. *Chattanooga News-Free Press*, July 22, 4.

Scopes, John T., and James Presley. 1967. *Center of the Storm: Memoirs of John T. Scopes*. New York: Holt, Rinehart and Winston, 41, 56.

Rhea County Courthouse
Est. 1891

The Rhea County Courthouse in Dayton, Tennessee, opened in 1892 after Dayton was named the seat of Rhea County. Today, the Romanesque Revival/Italian Villa–style building stands near the center of town, where it is bounded by Second and Third Avenues, and by Market and Court Streets. The building was designed by the Knoxville architectural firm W. Chamberlain and built by contractors from Chattanooga. The courthouse and its surrounding lawn cover four acres. The original dais on the courthouse lawn where Darrow confronted Bryan is gone, and the smaller one there now was built in 2005 for the History Channel's *Ten Days That Unexpectedly Changed America* (2006).

The three-story, brick courthouse, which includes a taller square tower, is famous for hosting the Scopes Trial on July 10–21, 1925. Most of the trial occurred in the second-floor's 65' × 65' (4,225 ft^2, with a 25'-high ceiling) courtroom, which Clarence Darrow described as "the largest courtroom that I have ever seen." From the judge's bench facing the audience, the prosecution sat to the left, the defense to the right, and the jury to the right, diagonally from the middle of the bench. Several reporters sat at tables to the left, between the prosecution and the judge. On an average day, about 600–800 spectators crowded the courtroom.

The courthouse was added to the National Register of Historic Places in 1972 and was designated a National Historic Landmark in 1976. In 1979, a $1 million project restored the second-floor courtroom to the way it looked during Scopes' trial.

The basement of the courthouse includes the Rhea County Heritage and Scopes Trial Museum, which exhibits artifacts such as trial records, photographs, and a WGN radio microphone used to broadcast the trial. Exhibits document what happened during the trial, as well as the trial's origins and aftermath. In July of most years since 1988, the trial has been reenacted in the courtroom as part of the Scopes Trial Play and Festival.

In 1955, the Tennessee Historical Commission placed marker (number 2B 23) titled "THE SCOPES TRIAL" in front of the courthouse to commemorate the significance of the trial. The marker reads as follows:

> Here, from July 10 to 21, 1925, John
> Thomas Scopes, a County High School
> teacher, was tried for expounding
> the theory of the simian descent
> of man, in violation of a lately
> passed state law. William Jennings
> Bryan assisted the prosecution:
> Clarence Darrow, Arthur Garfield
> Hayes and Dudley Field Malone
> the defense. Scopes was convicted.

This marker included errors; for example, Arthur Garfield Hays's surname was misspelled ("Hayes" instead of "Hays") and the Butler Act said nothing about a "theory of the simian descent of man." At Scopes Trial Day in 1960, Scopes challenged the marker's wording (Ogden, 1960), but the marker remained unchanged until sometime after 1975, when it was replaced with a marker claiming that Scopes had taught "that a man descended from a lower order of animals," and later by the accurate marker that stands in front of the courthouse today (Moore, 2016, 2021a, 2021b):

> Here, from July 10 to 21, 1925 John
> Thomas Scopes, a County High School
> teacher, was tried for teaching that
> man descended from a lower order
> of animals in violation of a lately
> passed state law. William Jennings
> Bryan assisted the prosecution;
> Clarence Darrow, Arthur Garfield
> Hays, and Dudley Field Malone the
> defense. Scopes was convicted.

On October 1, 2005, a statue of William Jennings Bryan commissioned by Bryan College was dedicated on the courthouse lawn. The statue, which was sculpted by Cessna Decosimo to commemorate Bryan College's 75th anniversary, depicts Bryan in 1891, when the courthouse opened and when Bryan began his four-year congressional career.

On July 14, 2017, a statue of Clarence Darrow paid for by the Freedom from Religion Foundation was unveiled near the statue of Bryan. Sculptor Zenox Frudakis, who sculpted Darrow to look as he did during the Scopes Trial (Fausset, 2017), included this portion of the transcript on the statue's base:

> **Darrow:** "Do you think the Earth was made in six days?"
> **Bryan:** "Not six days of twenty-four hours."

On March 1, 2021, Rhea County General Session Court Judge Shannon Garrison held court in the Rhea County Courthouse for the last time. All subsequent hearings moved to the new Rhea County Justice Center.

Sources

Fausset, Richard. 2017. To a statue outside the court, objections. *New York Times*, July 15, 9.
Moore, Randy. 2016. *A Field Guide to the Scopes Trial*. Dayton, TN: Rhea County Historical and Genealogical Society, cover.
______. 2021a. Correcting the history of the Scopes Trial. *Dayton Herald-News*, July 18, 1–2.
______. 2021b. The most famous historical marker in Rhea County. *Rhea Notes*, 23 (3), 1–3.
Ogden, Warren. 1960. Scopes challenges wording of marker. *Knoxville News-Sentinel*, July 24, A8.

Rhea County Drive-In Theater
Est. 1949

The 220-car Rhea County Drive-In Theater opened in 1949 about 2.5 miles north of Dayton on Rhea County Highway. On July 21, 1960, the drive-in was packed with cars, lawn chairs, and picnic blankets for the U.S. premiere of Stanley Kramer's *Inherit the Wind*, whose showing culminated Scopes Trial Day that celebrated the 35th anniversary of the Scopes Trial. Before the movie started, John Scopes and other dignitaries addressed the crowd from atop the concession stand. For years after the trial, the drive-in continued to occasionally feature Kramer's movie.

The Rhea County Drive-In Theater stopped showing movies nightly in the 1980s, and by 1999 its screen was a billboard. The screen was torn down in 2008 (Wilson, 2020).

Source

Wilson, Dean. 2020. The Rhea County Drive-In gone but not forgotten. *Yesterday in Dayton*, March 7, accessed November 6, 2020, at http://yesterdayindayton.net/rhea-drive-in/

Rhea County Heritage and Scopes Trial Museum
Est. 1978

On July 21, 1960, during the festivities of Scopes Trial Day, Dayton mayor J.J. Rodgers announced plans for a Scopes Trial Museum. The first donations to fund the museum came from a "wishing well" placed on the lawn of the Rhea County Courthouse into which people threw money during the celebration (Creation of Scopes Museum, 1960). The Rhea County Heritage and Scopes Trial Museum opened in October 1978 in the basement of the historic Rhea County Courthouse.

On June 11, 2018, Tennessee governor Bill Haslam officiated the reopening of the renovated museum. The museum features exhibits and artifacts from John Scopes' famous trial, Bryan College, the Trail of Tears, African American history, and Rhea County's military, industrial, musical, and agricultural history.

Source

Creation of Scopes Museum announced for courthouse. 1960. *Dayton Herald*, July 21, 1.

Rice, Kelso M.
1896–1986

Kelso M. Rice was born in 1896 in eastern Tennessee, and in 1924 joined the police department in Chattanooga. The next year, he was one of four officers from Chattanooga who went to Dayton to help with security at the Scopes Trial. When Judge Raulston chose Rice as the court's bailiff, Rice learned to "sound off with impressive determination" as he announced in the crowded courtroom that "people, this is not a circus. There are not monkeys up here. This is a lawsuit, so let's have order" (Officer Rice has famed pen, 1945).

When William Jennings Bryan died in Dayton five days after the trial, Rice was part of the American Legion's honor guard that attended his body. He was also the "lone guard" on the funeral train that took Bryan's body to its burial in Virginia. During that journey, Bryan's widow Mary noted that Rice had "such a good Christian face and stood so manly and erect at the head of the casket.... You are the representative of the simple people who loved my husband. In return, he loved you with tender affection" (Herrick and Herrick, 1925).

On the day after Bryan's burial, Rice brought the flag that had covered Bryan's casket to the Bryan family, which was staying at Washington's Lafayette Hotel. When he arrived, the Bryan's children were disputing who should get their late father's silver fountain pen, which he had used for 17 years for his correspondence and to sign the Versailles Peace Treaty (1919), Declaration of War with Germany (1917), Mexican Peace

Chattanooga police officer Kelso Rice (top row, center, with badge) was the court bailiff and part of the honor guard that attended Bryan's body after Bryan died in Dayton five days after the trial. This photograph was taken on the front steps of the F.R. Rogers home, where Bryan died (RCHGS).

Treaty (1917), and other documents. Mary settled the issue by giving the pen to Rice who, decades later, considered the pen "his most prized possession." Mary left her late husband's bedroom "as he left it," except for adding a memorial sent by an Iowa Sunday School class and a photo of Rice. As Mary told Rice, "you are still on guard in [his] room, and we are pleased to have you there" (Officer Rice has famed pen, 1945).

Rice spent the rest of his career in law enforcement; he became a squire (i.e., justice of the peace) and in 1957 was promoted to captain in the Chattanooga police force. He retired on November 1, 1961, and died at age 90 in Tampa Bay, Florida, on August 8, 1986.

Sources

Herrick, Genevieve Forbes, and John Origen Herrick. 1925. *The Life of William Jennings Bryan*. Chicago: Buxton Publishing House, 424.

Officer Rice has famed pen used in "Versailles" signing. 1945. *Chattanooga Daily Times*, August 26, 16.

Riley, William Bell
1861–1947

William Bell Riley was born on March 22, 1861, in Greene County, Indiana, less than a month before the outbreak of the Civil War. He was educated in nearby Valparaiso Normal School and—after receiving his teacher's certificate—graduated in 1885 from Indiana's Hanover College, ranking first in his class in debating. Three years later, Riley graduated from Southern Baptist Theological Seminary in Louisville, Kentucky, after which he pastored Baptist churches in Kentucky, Indiana, and Illinois. On March 1, 1897, Riley became pastor at blue-blooded First Baptist Church in Minneapolis, Minnesota, where he would spend the rest of his career. (When Riley arrived, the church's rich members rented the most visible pews.) Within five years, the church was a fundamentalist stronghold as it grew from 585 to more than 3,550 members. Riley also founded three "Northwestern Schools"—Northwestern Bible and Missionary Training School (1902), for pastors of rural churches; Northwestern Evangelical Seminary (1938), for pastors of urban churches; and Northwestern College (1944), for a Christ-centered liberal-arts education.

In May 1919 in Philadelphia, the tall, austere Riley founded what later became known as the World's Christian Fundamentals Association (WCFA). During his keynote address at the organization's first meeting, which attracted more than 6,000 attendees, Riley promised that his interdenominational organization would "bring the world back to the Bible" in an event of "more historic moment than the nailing up, at Wittenberg, of Martin Luther's ninety-five Theses" (Riley, 1919). Riley opposed gambling, prostitution, and drinking, but he blamed the "putrid condition" of America—including lewd theatre, dancing, and gambling—on the teaching of evolution (Riley, 1932). Riley was convinced that the teaching of evolution would damage morality and produce anarchy because "the whole vulgar theory tempts the virtuous to fresh vulgarities." To combat evolution, Riley promised that his WCFA—an organization "indisputably from God"—would espouse the "true gospel" and expose the "utter inharmony between evolution and the Christian faith" (Riley, 1926a; Trollinger, 1995).[39]

Riley used the "evolution question" to transform fundamentalism into a political movement. He was especially concerned about "pagan schools," claiming, "Every

In the 1920s, William Bell Riley—pastor of First Baptist Church in Minneapolis, Minnesota—organized millions of fundamentalists into the World's Christian Fundamentals Association. Bryan represented Riley's organization at the Scopes Trial in Dayton (First Baptist Church, Minneapolis, Minnesota).

Fundamentalist Christian ... is fighting a battle with the college next door" (Lienesch, 2012). By the early 1920s, he had focused the WCFA on driving the teaching of evolution from public schools, barnstorming the country to promote his cause. In October 1922, Riley—after proclaiming that "evolution is not a science, it is a hypothesis only, a speculation"—hosted a visit by William Jennings Bryan, who told nearly 12,000 people that "evolution is a menace to civilization" (Riley, 1922; Szasz, 1969). Bryan's speech inspired Riley to form the Antievolution League of America in 1924 "to force the teaching of the unproven evolutionary hypothesis ... from the tax-supported schools" (The Anti-Evolution League, 1923; Trollinger, 1995). In his many debates with scientists, Riley often proclaimed that perhaps "the Darwinians have vestigial remains in their bodies, but I want to say right now that I have none in mine." Bryan promised, "Within five years, not an educator in America will dare to look you in the eyes and say that he believes in evolution" (2800 of 3000 vote for Genesis, 1925; Shipley, 1927).

At its May 1925 meeting in Memphis, Riley's WCFA praised Tennessee for its antievolution law and, on May 12, asked Bryan to help Tennessee prosecute Scopes. The next day, from Pittsburgh, Bryan agreed to join the prosecution because "we cannot afford to have a system of education that destroys the religious faith of our children." Bryan, his era's best orator and most popular lay-Christian, added instant energy, recognition, and credibility to the WCFA and Riley's cause (Bryan will defend anti-evolution law, 1925). Although the most vocal fundamentalists were men such as Riley, Frank Norris, Billy Sunday, and John Roach Straton, one reporter estimated that, at the time of the Scopes Trial, 70 percent of all fundamentalists were women (Hartt, 1925).

As the Scopes Trial drew near, Bryan begged Riley to come to Dayton, but Riley instead went to Seattle to fight a takeover of the Northern Baptist Convention by modernists. From there, Riley denounced the Scopes Trial's biased coverage by evolution-promoting journalists who were "disgusting blood-suckers," and proclaimed that Bryan had won a "*signal conquest*" (emphasis by Riley) that had impressed judges, jurors, and "an intelligent world" (Trollinger, 1995). When Bryan died, Riley praised him profusely, but often added that Bryan was the fundamentalists' "unofficial" leader.

After the Scopes Trial, Riley stepped up his attacks, denouncing "unproven" evolution as "the evil seed" that produces atheism, "social slavery," and social putridity. At the WCFA's annual meeting in Toronto in May 1926, Riley announced plans to open fundamentalist colleges in every state. To Riley, science had become "the subtle word of Satanic employment," and modernist professors were "the greatest menace to Christianity and to American democracy" (Riley, 1926). Riley was especially concerned about the teaching of evolution at the University of Minnesota, where he pleaded that the school "not become the personal property of a dozen regents or a hundred Darwinized or Germanized, deceived, and faithless professors" (Russell, 1972; Riley assails "U," 1926). Riley, like most other fundamentalists of his era, rejected James Ussher's claim that Earth is only 6,000 years old because, according to Riley, "the Bible never taught any such thing." Instead, Riley—convinced that the Scopes Trial "has affected religious and educational life as nothing else in the last hundred years"—preached about the evils of evolution to overflow crowds at his church and elsewhere, in one span denouncing evolution from his pulpit on 10 consecutive Sundays (Riley, 1927; Russell, 1972). Each of his speeches and sermons about evolution were heard by thousands of people.

On February 25, 1927, at Riley's urging, Minnesota state senator K.K. Solberg introduced a bill to ban the teaching of evolution in Minnesota's public schools. Two weeks later, Riley pleaded his case in a joint session of the state legislature, but it made little difference. On March 10, *The Minnesota Daily* delivered the results of the vote in its front-page headline: "Senate Buries Evolution Bill for Four Years by 55-to-7 Vote." Undaunted, Riley dismissed the defeat as "only a skirmish" and, at its annual convention two months later in Atlanta, announced, "We have arranged the machinery for a finish fight with the backers of evolution to mold sentiment against the evolutionary hypothesis and to get voters educated to the point they will express themselves through the ballot as the Fundamentalists wish" (Bailey, 1953).

Riley was wrong; his bill in Minnesota was never reintroduced and, except for an antievolution law passed by popular vote in 1928 in Arkansas, his 1927 campaign in Minnesota was the WCFA's last major effort to ban the teaching of evolution. In the 1930s, Riley continued to publish antievolution pamphlets—for example, *Are the Scriptures Scientific?*, *Darwin's Philosophy and the Flood*, and *Evolution, A False Philosophy*—but his cause was lost. Riley later blamed a "Jewish-Bolshevik-Darwinist conspiracy" for his failed antievolution campaign (Trollinger, 1990, 1995).

After peaking at nearly six million members in 1927, Riley's WCFA began to decline, as did attendance and donations at Riley's church. Riley resigned as WCFA's president in 1929, after which financial troubles ended the WCFA's antievolution campaign. Although Riley continued to denounce evolution until he retired in 1942, the public tired of the controversy. As Curtis Akenson, a subsequent pastor of Riley's church noted, "Riley's early preaching about evolution filled the church. Riley's later preaching about evolution emptied the church" (Hughes and Moore, 2015). Historian Ferenc Szasz described Riley as "a social radical when he arrived in Minneapolis in 1897, on the left, and a social radical when he died in 1947, but on the right. Riley went from radical to radical by standing still. For fifty years he stood on the corner and the parade passed him by" (Szasz, 1969). Nevertheless, even at age 84, Riley celebrated his debate victories over evolutionists, claiming that "I sincerely regret that [evolution's] advocates decided to abandon the field of debate; I cannot blame them" (Riley, 1945).

On his deathbed, Riley chose Billy Graham to succeed him as president of Northwestern.[40] Graham, against his better judgment, accepted the job, in which he served for 3.5 years. Graham often praised Riley, but later admitted, "I learned a great deal from some of the mistakes I thought he had made" (Russell, 1972).

On December 5, 1947, Riley—"The Grand Old Man of Fundamentalism"—died at his home in Golden Valley, Minnesota. He is buried in Lakewood Cemetery in Minneapolis. Riley, a militant antimodernist who wrote more than 60 books and hundreds of articles, was the most influential fundamentalist of his generation.

Sources

Bailey, Kenneth Kyle. 1953. The Anti-Evolution Crusade of the Nineteen-Twenties. Ph.D. Dissertation, Vanderbilt University, Nashville, Tennessee, 53.
Bryan will defend anti-evolution law. 1925. *New York Times*, May 13, 15.
Hartt, Rollin Lynde. 1925. What lies beyond Dayton. *Nation*, 121, 111.
Hughes, Andrew, and Randy Moore. 2015. Measuring William Bell Riley's anti-evolution crusade in Minnesota. *Reports of the National Center for Science Education*, 35, 1.1–1.13.
Lienesch, Michael. 2012. Abandoning evolution. *Isis*, 103, 693.
Riley assails "U" as fostering state atheism. 1926. *Minneapolis Tribune*, March 8, 1.
Riley, William Bell. 1922. The evolution controversy! *Christian Fundamentals in Schools and Church*, 4 (April–June), 5.
______. 1926. *Inspiration or Evolution*. Cleveland: Union Gospel Press.
______. 1927. Mr. Bryan and the Scopes Trial. 1927. *Christian Fundamentals in School and Church*, 9 (January–March), 9.
______. 1945. The sickly birth of evolution. *The Northwestern Pilot*, 26 (October), 3.
Russell, C. Allyn. 1972. William Bell Riley: Architect of Fundamentalism. *Minnesota History*, Spring, 14–30.
Shipley, Maynard. 1927. *The War on Modern Science: A Short History of the Fundamentalists' Attacks on Evolution and Modernism*. New York: Alfred A. Knopf, 352, 371.
Szasz, Ferenc M. 1969. *Three Fundamentalist Leaders: The Roles of William Bell Riley, John Roach Straton, and William Jennings Bryan in the Fundamentalist-Modernist Controversy*. Doctoral Thesis, University of Rochester.
Trollinger, William Vance. 1990. *God's Empire: William Bell Riley and Midwestern Fundamentalism*. Madison: University of Wisconsin Press.
______. 1995. *The Anti-Evolution Pamphlets of William Bell Riley*. New York: Garland, xix.
2800 of 3000 vote for Genesis. 1925. *Spokesman-Review* (Spokane, WA), June 25, 1.

Robinson, Frank Earle "F.E."
1881–1957

Frank Earle "F.E." (often misidentified as "Fred") Robinson was born in Evensville, Tennessee, in June 1881. Five years later, Robinson's family moved to Dayton, where F.E. studied at the Masonic School. He then started a laundry business and used its profits to open, with partner John Crawford, the "Crawford and Robinson" drugstore in 1909. Three years later, Robinson—by then a registered pharmacist—bought out Crawford and changed the name of the business to F.E. Robinson Company. This store on Main Street near Market Street became Robinson's Drug Store, which was a gathering place and cultural center in Dayton (Broyles, 1991). (Robinson also owned a drugstore in nearby Spring City, which was operated for years by his son, W.C. "Sonny" Robinson.) Robinson also sold textbooks for the local schools, including those used by John Scopes.

In 1910, Robinson married Clark Eliza Haggard, the daughter of A.P. and Rena (Clark) Haggard. As a wedding gift, A.P. Haggard gave the newlyweds a home at the northeastern corner of Market Street and Third Avenue. As of 2022, that home was still owned by descendants of the Robinsons.

On May 5, 1925, Robinson—with local engineer George Rappleyea, Rhea County School Superintendent Walter White, and White's assistant Clay Green—instigated the Scopes Trial at a small table in Robinson's Drug Store. After Scopes agreed to be prosecuted, Robinson broke the news to the *Chattanooga News*. Robinson and other civic leaders then began raising $5,000 to "tell the world about Dayton." Throughout Scopes' trial, Robinson entertained visitors at his drugstore, led tours of Dayton, and stretched a large banner across Main Street to tell people that his store was "Where It Started." The Underwood & Underwood photographers that Robinson hired to document the trial took many of the most popular photographs of the trial.

Robinson's wife, Clark, wrote articles (without credit) for area newspapers and was F.E.'s publicist. She did most of the work to produce *Why Dayton—of All Places?*, a 28-page booklet (coauthored with W.E. "Bill" Morgan) extolling the virtues and commercial potential of Dayton as a place for "champions of the survival of the fittest" and "you followers of the lowly Nazarene."

On Wednesday, July 15, 1925, Robinson—chairperson of the school board—testified in Scopes' trial that the state-approved textbook (i.e., George Hunter's *A Civic Biology*) used by Scopes earlier in the year had been used in Tennessee since 1919. Robinson also corroborated school superintendent Walter White's testimony that Scopes had admitted that "any teacher in the state who was teaching Hunter's Biology was violating the law." When Clarence Darrow asked if Robinson sold Scopes' textbook, Robinson responded, "Yes, sir," at which time Attorney General Tom Stewart noted that the Butler Act criminalized the teaching of evolution, not Robinson's selling of textbooks that included evolution. When prompted by Darrow, Robinson read the first paragraph on page 194 of Hunter's book: "The Doctrine of Evolution.—We have now learned that animal forms may be arranged so as to begin with very simple one-celled forms…." Robinson was the fourth and final witness called by the prosecution. After the Scopes Trial, Robinson continued to promote his drugstore's involvement in the trial. Later in her life, Robinson's daughter, Frances Robinson Gabbert, recalled that the Scopes Trial "was fun. I think everyone would agree about that. But I don't know that it could ever happen again" (George, 1985).

On November 18, 1932, Robinson was indicted for violating banking laws in connection with Dayton's then-closed American National Bank. Robinson's alleged co-conspirators were A.B. Andrews, a former official of the Dayton Hosiery Mills; A.P. Haggard, a prominent civic leader and president of the bank; and Wallace C. Haggard, A.P.'s son and a bank employee. After the elder Haggard committed suicide at Robinson's home, all of the charges were dropped (Tennessee bank case, 1932).

Robinson, who worked as a druggist in Dayton for half a century, was one of Dayton's most widely known civic, educational, and business leaders. He chaired many of Bryan College's original major committees, and for 25 years led the board of trustees. At one point, a large neon sign in Robinson's yard pointed visitors to Bryan College. Robinson's home is marked with a "Scopes Trial Trail" plaque, and Bryan College's Robinson Hall dormitory (dedicated March 19, 2007) honors his service as an "entrepreneur, pharmacist, civic leader, Scopes Trial planner and friend of William Jennings Bryan, incorporator and president of the Bryan Memorial University Association … incorporator of William Jennings Bryan University, founder [with wife Clark Haggard Robinson] of

William Jennings Bryan University, and first chairman of the Board of Trustees" from 1930 to 1956.

For decades, Robinson's Drug Store displayed the table at which the Scopes Trial originated, marking it with a sign saying, "At this table the Scopes Evolution Case was started May 5, 1925." In April 1957, when Robinson sent the table to Chicago to be displayed at the Clarence Darrow Centennial celebration, "people would come in and ask where the Scopes table was" (Scopes Trial figures recall old memories, 1960). Days later, on Friday, May 3, 1957, the 75-year-old Robinson died of a heart attack in his home. Following a funeral service at the Robinson home on the following Sunday afternoon, Robinson was buried in Dayton's Buttram Cemetery (F.E. Robinson, 1957). Robinson's death was front-page news throughout eastern Tennessee (Dayton druggist in Scopes case dies, 1957).

After Robinson's death, the operation of the Dayton drugstore transferred to F.E.'s son, Sonny. Robinson operated the store until 1982, when declining health left him unable to staff the store. That year, Robinson closed the store and moved its remaining supplies, furniture, and inventory to the Robinsons' store in Spring City, which was operated by Sonny's wife, Katherine.

Sources

Broyles, Bettye J. 1991. *History of Rhea County, Tennessee*. Dayton, TN: Rhea County Historical and Genealogical Society, 353.
Dayton druggist in Scopes case dies. 1957. *Knoxville News-Sentinel*, May 3, 1.
F.E. Robinson, 75, dead at Dayton. 1957. *Chattanooga Times*, May 3, 1.
George, Dan. 1985. The day the circus came to town. 1985. *The Jackson (Tennessee) Sun*, July 21, 9.
Scopes trial figures recall old memories. 1960. *Knoxville News-Sentinel*, July 17, B-10.
Tennessee bank case brings 4 indictments. 1932. *New York Times*, November 19, 8.

Robinson's Drug Store
Est. 1909

Robinson's Drug Store opened (originally as "Crawford and Robinson") on March 18, 1909, on Dayton's Main Street, just west of Market Street and adjacent to Hotel Aqua. John Crawford later sold most of the store's operations to his partner Frank Earle Robinson, who made the store an integral part of Dayton's business and social life. Robinson's Drug Store sold prescriptions, sodas, magazines, and textbooks used by students at Rhea Central High School, including *A Civic Biology* from which John Scopes taught in April 1925 and, in the process, allegedly violated the Butler Act.

On May 5, 1925, a small group of Dayton businessmen instigated the Scopes Trial at a table in Robinson's Drug Store in hopes of improving Dayton's struggling economy. John Scopes was summoned from a tennis match, and the businessmen convinced Scopes to test the law. Throughout the trial, Robinson's Drug Store was a gathering place that one reporter described as "the capital of the universe" (Lait, 1925). For decades afterward, Robinson displayed the table at which the businessmen sat with a sign: "At this table the Scopes Evolution Case was started May 5, 1925." Residents and tourists alike wanted to see the table (Ogden, 1960).

During the Scopes Trial, Bryan often cooled himself in the sweltering courtroom with a hand-fan that advertised Robinson's Drug Store. Meanwhile, the drugstore

The Scopes Trial was hatched by a group of businessmen at Robinson's Drug Store in downtown Dayton. During and after the trial, Robinson advertised his store as the place "Where It Started" (RCHGS).

bustled with activity; Bryan, Darrow, Scopes, and others visited the store for drinks and conversation. To accommodate the crowds, Robinson—for the first time ever—opened his store on Sundays during the trial. Robinson's most popular drinks were Cokes (5¢), "Cantaloupe Sundae" (25¢), and a "Bryan Fizz" (10¢) made from ginger ale and "secret ingredients" (Vignettes and tidbits, 1981). Business was good; he sold 1,280 lemonades, nearly the same number of limeades, 768 chocolate drinks, 760 Coca-Colas, and 20 gallons of ice cream per day during the trial's first week (Some drinkers in Dayton town, 1925). Robinson also hosted tours and distributed the booklet *Why Dayton—of All Places?* to lure families and investors to Dayton. For several weeks before, during, and after the trial, a banner stretched across Main Street in front of Robinson's telling people his store was "Where It Started."

By 1935, Robinson had moved his drugstore to Market Street, where the upstairs rooms became a hospital. In April 1957, Robinson loaned his store's famous table to Chicago for display in the Clarence Darrow Centennial celebration. In 2019, the Robinson family donated that table and its accompanying chairs to the Tennessee State Museum (Table and chairs, 2019). A replica is sometimes displayed in Dayton's Scopes Trial Museum. When Scopes returned to Dayton in July 1960 for Scopes Trial Day, he and his wife Mildred visited with many of John's friends at Robinson's Drug Store.

Robinson's Drug Store closed in 1982, at which time the remaining furniture and supplies were transferred to Robinson's store in Spring City. Although the building on Main Street that housed the original Robinson's Drug Store was torn down in 1989, a "Scopes Trial Trail" plaque stands near its former location. In 2014, Dayton honored the store's legacy by marketing a Christmas ornament depicting the store in 1925.

Sources

Lait, Jack. 1925. Dayton fears Scopes Trial prove a dud. *Messenger-Inquirer* (Owensboro, KY), July 12, 11.
Ogden, Warner. 1960. Scopes Trial figures recall old memories. *Knoxville News-Sentinel*, July 17, B-10.
Some drinkers in Dayton town. 1925. *Chattanooga Daily Times*, July 18, 11.
Table and chairs from Robinson's Drug Store where the Scopes Trial was planned. 2019. *Tennessee State Museum Newsletter*, Summer, 14–15.
Vignettes and tidbits of the trial. 1981. *Chattanooga Times*, March 14, 1.

Rogers, Frederick Richard 1895–1964

Frederick Richard "F.R." Rogers was born in Rhea County on May 16, 1895. After serving in World War I and graduating from the Philadelphia College of Pharmacy, Rogers returned to Dayton. He was working as a pharmacist at Robinson's Drug Store at the time of the Scopes Trial.

When Bryan and his entourage arrived in Dayton in early July 1925, they rented Rogers's home for $25 per week. (Rogers and his family moved to their cabin on the nearby mountain.) During the trial, Bryan often walked around Dayton, where he visited with local residents and bought radishes for snacks.

After the trial, Bryan gave several speeches in surrounding communities. On the morning of Sunday, July 26, Bryan returned to Dayton, where he appeared unannounced at First United Methodist Church (South) and led a prayer. Bryan then went to the Rogers's home for lunch. Known for eating hearty meals, Bryan ate two chickens and what the cook called "all the trimmings," and then laid down for an afternoon nap. When William "Jimmy" McCartney—Bryans' chauffeur and assistant—could not rouse him, the Rogers's cook ran next door to the A.B. Andrews's home to get help. A.B. Andrews then ran across the street to summon Walter Thomison, a physician, but he could not revive Bryan. Bryan died in the Rogers's home (Guffey, 2018). Prosecutors Sue Hicks, Wallace Haggard, Herbert Hicks, and Gordon McKenzie watched Bryan's body through the night.

Bryan's body laid in repose in the Rogers's parlor on July 27–28, where it was viewed by hundreds of mourners. Among the first arrivals was George Rappleyea, who broke the story of Bryan's death with the Universal News Service. On July 28, Dayton preacher Charles Jones led memorial service for Bryan from the Rogers's front porch. Bryan's body was then taken by train to Washington, D.C., where his funeral was held at New York Avenue Presbyterian Church. He was buried in Arlington National Cemetery in Arlington, Virginia, beneath the epitaph "He Kept the Faith."

In 1926, Rogers and Ed Pierce opened the Rogers and Pierce Drug Store at the intersection of Main and Market Streets. A few years later, Rogers bought out Pierce and the store became the F.R. Rogers Pharmacy. Above Rogers's pharmacy was the 18-room Thomison Hospital (Broyles, 1991).

Rogers helped found Bryan College and was a trustee of the college from near its inception to his death. He died of a heart attack at a dinner at Bryan College on November 20, 1964, and is buried in Dayton's Buttram Cemetery. Rogers's house was later destroyed, and a new house was built on the land; this site, at the intersection of Market Street and Georgia Avenue, is marked by a "Scopes Trial Trail" plaque. Rogers's pharmacy closed in 2004. A stained-glass window at First United Methodist Church (South) honors Rogers's many years of service to the church.

Sources

Broyles, Bettye J. 1991. *History of Rhea County, Tennessee*. Dayton, TN: Rhea County Historical and Genealogical Society, 353.
Guffey, Pat Hawkins. 2018. The last meal and demise of William Jennings Bryan. *The Herald-News* (Dayton, Tennessee), July 13, 4.

Ryan, Quin Augustus 1898–1978

Quin Augustus Ryan was born on November 17, 1898, in Chicago, Illinois. Quin attended Loyola Academy, after which he enrolled at Northwestern University, where he was active in writing and theater. He began his journalism career as a sports reporter, photographer, and editor with *The Chicago Tribune*, and soon thereafter was working for WDAP radio. In 1924, the *Tribune* bought WDAP and changed its call-letters to WGN, which was derived from the *Tribune*'s slogan, "World's Greatest Newspaper." Ryan became a pioneer in radio broadcasting, anchoring the first live broadcasts of a circus, Kentucky Derby, Indianapolis 500, a baseball game, and other events that were "almost as good as being there." In 1924, WGN had planned to broadcast the entire Leopold and Loeb murder trial, but decided that its gruesome nature made that unwise. Then, according to Ryan, "here comes this big, clawing thing [i.e., the Scopes Trial] down in Tennessee, and everybody is making fun of it, so that's for us.... Good entertainment" (Wesolowski, 1975).

In July 1925, Ryan—assisted by engineer Paul Neal and Scopes' student Carmack Waterhouse—anchored WGN's broadcast of the Scopes Trial. Ryan had hoped to begin broadcasting when the trial opened on July 10, but technical problems delayed live coverage until the following Monday (i.e., July 13). During WGN's broadcasts, Ryan spent much of his time sitting quietly with journalist H.L. Mencken in a windowsill, functioning as a "pointer" who clarified to listeners the issues being raised in court and identified the people who were speaking. Ryan, who earned $60 per week for his work, was WGN's only announcer in Dayton. Ryan's broadcasts from 10:00 a.m.-noon and from 2:00–4:00 p.m. were heard throughout the Midwest, but people on the east and west coasts probably could not hear them. One evening, in hopes of attracting a nighttime audience, Ryan aired interviews and "special speeches" from Darrow, Bryan, Scopes, Rappleyea, and Raulston while pretending that he was walking along the street. During the trial, Ryan considered himself one of the "mods, we were the sophisticates, we were the intellectuals. And they were the hillbillies.... If I showed any partiality, it was just because the newspapermen all did ... we were all plugging for John Scopes..." (Wesolowski, 1975). According to Ryan, none of the attorneys paid much attention to WGN's courtroom microphones.

Although most journalists left town before the trial's final two days, Ryan and Neal broadcast from Dayton until the trial was over. It was compelling radio, but because there was no technology to record the broadcasts, there are no recordings of Ryan's broadcasts from Dayton (Wesolowski, 1975).

Ryan later broadcast other prominent national events, including the 1928 Democratic and Republican Conventions, the Dempsey-Tunney fights, and Knute Rockne's funeral. Ryan was a pioneering announcer; no one had done such events live before (Currie, 1976). By 1931, Ryan managed WGN, and in 1941 he left WGN to do freelance writing and consulting. In 1947, the ABC radio network aired a reenactment of the Scopes Trial's highlights that featured Ryan, but this broadcast, like the original trial, was not recorded.

Ryan died at Chicago's Whitehall Nursing Home on October 7, 1978, and is buried at Calvary Cemetery in Evanston, Illinois.

Sources

Currie, William. 1976. Quin Ryan, dies at 79, Chicago radio pioneer. *Chicago Tribune*, October 8, 1978, Section 3, 23.

Wesolowski, James Walter. 1975. Before Canon 35: WGN broadcasts the Monkey Trial. *Tennessee Historical Quarterly*, 34 (4), 392–406.

Salem, Illinois Est. 1823

William Jennings Bryan (1860–1925), the most famous of John Scopes' prosecutors at the Scopes Trial, was born in Salem, Illinois, on March 19, 1860. Bryan attended Salem Academy and was a member of Salem's Cumberland Presbyterian Church (now First United Presbyterian Church); during a visit to the church later in his life, Bryan donated an ornate podium, which continues to stand in the church sanctuary. Throughout his life, Bryan returned often to Salem to give speeches, several of which were heard by John Scopes and his family.

When John Scopes' family moved to Salem in 1916, they attended Cumberland Presbyterian Church. The family lived in Badollet House, which had been a stop on the Underground Railroad and the first brick home in Salem. When Scopes graduated from Salem High School on May 16, 1919, Bryan was the commencement speaker at Salem's Methodist Episcopal Church (today's Grace United Methodist Church). At that ceremony, Scopes and three of his classmates interrupted Bryan's talk with "boyish frivolity" and laughter, which both Scopes and Bryan remembered when they met in Dayton six years later.

Bryan's birthplace, Badollet House, and Salem's Methodist Episcopal Church are both on the National Register of Historic Places. The William Jennings Bryan Birthplace Museum includes photos, newspaper clippings, awards, books (e.g., *The Menace of Darwinism*), life-masks of Bryan and his wife Mary (created by Gutzon Borglum, the sculptor of the memorials at Mt. Rushmore), and other artifacts of Bryan's life. The Marion County Courthouse on Salem's town square features a mural depicting Bryan's childhood.

Across from Salem's Bryan Memorial Park stands a large, bronze statue of Bryan. This statue, which was sculpted by Borglum, was dedicated in West Potomac Park in Washington, D.C., by President Franklin Roosevelt on May 3, 1934. It was later moved to accommodate the Theodore Roosevelt Bridge, which opened in 1964. After laying in a vacant lot for several years, the 2,700-pound, 8'-high statue and its 29,300-pound marble base were moved in 1961 to Salem, where they now stand. (Scopes was invited to the dedication, but he got sick on the way to Salem and did not attend.) The transfer of Bryan's statue to Salem was the first time a Washington Park Service statue had ever left the Capital. Salem also hosts WJBD radio, whose first three call-letters (i.e., "WJB") honor

In 1934, this statue of William Jennings Bryan was placed on the National Mall in Washington, D.C. The statue was moved in 1958 during construction of the Theodore Roosevelt Bridge, and in 1961 it was moved to this site in Salem, Illinois, Bryan's hometown. This statue was sculpted by Gutzon Borglum, who is best known for his work on Mount Rushmore and Stone Mountain (Randy Moore).

Bryan. Salem's *Times-Commoner* newspaper also honors Bryan, who was often called "The Commoner" and published his popular *The Commoner* newspaper from 1901 to 1923.

When Bryan's friend and supporter Philo Bennett died in 1903, his estate included $1,500 for a library to be built where Bryan was born. Bryan matched Bennett's contribution and the Bryan-Bennett Library opened in 1909. The library's first books were bought with money donated by Bryan.

Source

Moore, Randy. 2019. William Jennings Bryan and John Scopes in Salem, Illinois. *Reports of the NCSE*, 39 (4), 10.

Scopes, John Thomas
1900–1970

John Thomas Scopes was born on August 3, 1900, in Paducah, Kentucky, the last of six children born to Thomas and Mary Alva Brown Scopes (Appendix 1). John was born in the Scopes' home at 4 Mayfield Gravel Road, which was just across from the railway yard where Thomas worked. John described his childhood home as "a tolerant environment that taught me early in life to revere truth and love and courage." Family members and friends referred to John as "John Thomas" and "J.T."

The Thomas and Mary Alva Scopes' family, circa 1911. Back row, left to right: Lela V Scopes, Thomas Scopes, Mary Alva Scopes, Irene Scopes, Martha Elisabeth Calvert Brown (Mary Alva's mother), and Ethel Scopes. Front row, left to right: Harold Gilliam, John Thomas Scopes, Nannie Mae Scopes Gilliam, and her baby, Nannie Mae Gilliam (Jeanette Gilliam Travis).

In 1913, the Scopes family moved to Danville, Illinois, where John and his sister Lela were "considered Southerners [and] placed at the front of the classroom with the two African-American kids in the classroom" (Scopes and Presley, 1967). As John was completing the ninth grade, the family relocated to Salem, Illinois, where John played forward on the high school basketball team. John was a "very studious" student who "liked a good argument. He wanted to argue everything out" (Montgomery, 1973). When John graduated from Salem High School in Spring 1919, his commencement speaker was hometown hero William Jennings Bryan, who John would see again at his trial in Dayton.[41]

In the Fall of 1919, after spending the summer in Kentucky shoveling coal, John enrolled as a chemistry major at the University of Illinois at Urbana, but he quit after one year because of bronchitis and went home to recuperate. After he "hoboed around for a while," John enrolled in the Fall of 1921 at the University of Kentucky, where he earned an A.B. degree in arts-law in Spring 1924. After graduating, while at home with his parents in Paducah, sister Lela helped John apply for several teaching jobs, and he received two offers. He chose the lesser-paying job as a coach and science/math teacher in Dayton, Tennessee, because Dayton was "a small town with a small school where I [wouldn't] get into any deep water" (Shelton and Smith, 1979).

By late summer, 1924, Scopes was working at Dayton's Rhea Central High School (RCHS), where he taught science/math and coached football, basketball, and baseball. Scopes soon realized that football in Dayton "was more important than bootlegging" and that coaching "was the most important part of my job" (Football, evolution, and bootlegging, 1925; Scopes and Presley, 1967). This must have been daunting to Scopes, for—except for basketball in high school—he had never participated in organized sports.

Coach Scopes

Dayton was anxious for a winning football team. In the Fall of 1922, two years before Scopes arrived in Dayton, the RCHS "Yellow Jackets" football squad had high hopes; Vanderbilt football star Wallace Haggard—who would help prosecute Scopes in 1925—had returned to Dayton to coach the team (Rhea High gridders, 1922). On September 22, despite an enthusiastic, record-breaking crowd, the Yellow Jackets—after defeating Rockwood 81 to 0 the year before—were held by Rockwood to a 0–0 tie. Thereafter things got worse, with the Yellow Jackets enduring a winless, scoreless season (Rhea's green team, 1923).

The following year, the Yellow Jackets had a new coach—Raleigh Reece, a former football star at Carson and Newman College (now Carson-Newman University). After posting a 5–6 record, Reece resigned to become a reporter for the *Nashville Banner*, and was replaced by John Scopes. The chain-smoking Scopes—the first coach to let his players smoke during the season—posted a 4–4–2 record, which prompted one reporter to announce, "Good football coaches have been scarce around here … there may never be another one like John Scopes," and another to declare Scopes "the best football coach that Dayton ever had" (Monkey war stirs Tennessee, 1925; Football, evolution, and bootlegging, 1925).

When Scopes left Dayton after his infamous trial, he was replaced as football coach by Raleigh Reece, his predecessor.[42] The day before he was hired, and presumably to help him get the job, Reece announced that he was a fundamentalist, prompting one RCHS

Throughout much of his early adult life, John Scopes often stayed with his parents in Paducah, Kentucky, where he was born. This photograph shows John on the front porch of his parents' home on Berger Road, circa June 1925 (Jeanette Gilliam Travis).

student to lament, "Shucks, our football team is busted.... What does a fundamentalist know about football?" (Darwin in Scopes' place, 1925). Reece's team had another losing season, winning four games and losing seven. Although the team's official name remained the "Yellow Jackets," several newspapers had, after the Scopes Trial, renamed the team "The Fundamentalists" (Rhea High Fundamentalists, 1926). Scopes never again coached a sports team, not even those that included his two sons.

When not coaching, Scopes taught science and math. By all accounts, the youthful Scopes was a popular teacher, and he often socialized with students at his church (Dayton's Cumberland Presbyterian Church) and Saturday night dances. According to one report, Scopes—described as looking like "a college student on vacation"—was "the hero of the students," who were "persuaded by his quiet demeanor" (Anderson, 1925). According to Scopes, the people of Dayton "liked me ... [and] I liked them" (John T. Scopes discusses, 1960).

In April 1925, William Ferguson—the RCHS principal and biology teacher—was sick, and Scopes was recruited to teach his biology class for two weeks. Scopes thought little about his alleged coverage of evolution, usually claiming that "I didn't remember teaching it" and "I didn't want to start any new subject, so I just reviewed what the class had been over" (e.g., Scopes and Presley, 1967). However, at other times, Scopes admitted that "I did cover the Darwin theory of evolution," adding, "I don't see how

I could have reviewed [biology] for 10 days and not touched on evolution" (Strongen, 1960; Ogden, 1960; Shannon, 1970). The best judges of whether Scopes had taught evolution were Scopes' students, who were certain that he *had* taught evolution—not just when substitute teaching in Ferguson's biology class, but also in his general science and physics classes (Of Monkeys and Men, 2021).

Scopes' Arrest and Trial

When classes ended on May 1, Scopes stayed in Dayton to check on two students who had been injured in a car wreck and to pursue "a beautiful blonde" who, as he later admitted, deserved "as much credit as anyone for the trial being held in Dayton that summer" (Olasky and Perry, 2005). On May 4, at Robinson's Drug Store, engineer George Rappleyea showed several other Dayton businessmen an article in the Chattanooga newspaper describing how the ACLU was looking for someone to test Tennessee's newly passed law banning the teaching of human evolution in the state's public schools. The group cared little about evolution, but hoped that a show-trial might stimulate the area's struggling economy. Scopes was summoned from a tennis match and asked if he would be the defendant in the case. Scopes agreed and was charged with violating the law. No one could imagine the furor that Scopes' case would produce.

After being arrested, Scopes often accommodated reporters, telling them that he accepted evolution "from one end to the other" but "didn't know that I'd started anything. I didn't know that I was violating the law" (Snook, 1925). He condemned the Butler Act as "a revival against science," adding, "They told me they were going to arrest me and I told them to go ahead ... only people who have had their emotions played upon believe science and religion conflict.... I am going to see [this] through. The only defense that I will make is that the law is unconstitutional ... the question in this case is one that will have to be decided some time and might as well be now" (Dayton, Tenn., put "on map," 1925; Other teachers equally guilty, 1925).

In early June, Scopes went to New York to consult with the ACLU and some of the attorneys who would be defending him. The ACLU resisted Clarence Darrow's involvement in the case, but Scopes insisted on having Darrow defend him, adding that he felt the case "demanded an Indian fighter rather than someone who had graduated from the proper military academy" (Scopes and Presley, 1967). While in New York, Scopes toured the American Museum of Natural History, where he met with famed paleontologist Henry Fairfield Osborn. Osborn, who Scopes' defender John Neal said "did more than any other scientist to aid us in the Scopes case" ("Battle of Dayton," 1927), urged Scopes to stress that evolution was compatible with the Bible. Scopes did just that, telling reporters, "I am a Christian. I believe the Bible was inspired and that Christ did exist" and "I do not see how [evolution] conflicts with the Bible. A person can believe in evolution and believe in the Bible, too" (Scopes, back home, 1925; Scopes tells of religious beliefs, 1925). On June 16, Scopes reapplied for his job at RCHS, as did scores of other people from across the country (Scopes applies for job, 1925; Many science teachers seek Scopes' position, 1925). Scopes later admitted to being "an agnostic, but ... devoutly religious in my own way" (Stone, 1941).

Uncomfortable with what he considered invasions of his privacy, Scopes became increasingly skeptical of reporters, "strange dames in town stalking him," and other people wanting to talk with him (Potter, 1951). In New York, unlike in small-town

Dayton, he was hounded by aggressive reporters everywhere he went. The onslaught began when he arrived at his hotel, where Scopes was overwhelmed by "upward of 50 reporters," who Scopes said took "just a few of my words and expanded them into a column" (Snyder, 1925). Scopes quickly learned that even when he refused reporters' questions, they invented stories, and his silence "didn't make any difference" (Milton, 1925). The frustrated Scopes complained, "There is nothing pretty about New York ... the only place I saw ... that interested me was the public library ... [I have] never seen anybody so greedy as the restaurants and hotel keepers ... [or] so selfish as the newspaper men who have been dogging my trail." In response, newspapers reported that Scopes "knows little about Darwinism" and was "ill at ease when stammering brief explanations" of his case (Garrett, 1925; Scopes ill at ease, 1925). Fed up with New York, Scopes admitted, "I'm a country boy, pure and simple" and promised that "as soon as this evolution business is over, I'm going back to Kentucky where I belong" (Snyder, 1925). Reporters didn't let up, claiming that Scopes—now "the most famous individual on the continent"—was finding "the hero business different from what it is cracked up to be" as he continued to "play hooky from the spotlight which shines for him" (Lait, 1925). Newspapers featured pictures of Scopes frolicking with women while claiming that he was "taking his conviction lightly," "he isn't worrying," "he doesn't have to swim alone," and being "as popular with the girls as with his football team" (Moutoux, 1925; Taking his conviction lightly, 1925; Scopes, awaiting trial, 1925). As Scopes noted later, "the pressure of publicity never let up" (Scopes and Presley, 1967).

Newspapers claimed that Scopes had rejected movie and publicity offers "aggregating $150,000." Scopes later explained that there were "no such offers," adding that his trial was "a serious matter" and "my work in life is to teach and I know nothing of acting. If I take money, I want to feel that I have earned it" (Bailey, 1953). Scopes knew that he "could have made some money if I wanted to make myself unhappy for three or four years," but he refused "to be an exhibit. Besides, I don't know enough about evolution to lecture on it" (Doyle, 1960; Ginger, 1958). As his trial approached, Scopes told reporters , "I haven't worried about [my trial] any, except that the affair has caused my mother some anxiety.... I hope to be vindicated.... This is a test suit and not a discussion of religion. When this thing started we had no idea it would ever get beyond Rhea County." Scopes then added, "The [Butler Act] is an act of intolerance, and if allowed to stand it will pave the way to more intolerance.... A great many condemn [evolution] who do not know anything about it" (Scopes, back home, 1925). A Baptist editor had a different view of Scopes' case, denouncing Scopes as "just a fool boy who has lent himself to be the tool of faddists and opportunists" (Baptist editor supports Bryan, 1925).

By the time his trial began, Scopes' sister Lela had been fired from her teaching job in Paducah when she would not renounce her now-famous brother's beliefs in evolution. Elsewhere, the media was focused on Bryan and Darrow, and John Scopes was largely forgotten; newspapers claimed that "Johnny Scopes" would contribute "incidental features" to the proceedings, and Scopes was left to sit "speechless, a ringside observer at my own trial, until the end of the circus" (Evolution trial opens this morning, 1925; Scopes and Presley, 1967). One afternoon, when Scopes returned to court late from a lunch break, proceedings began without him.

During the trial's second weekend, Scopes wrote his first trial-related article, telling readers around the world that trial instigator George Rappleyea deserved "the glory" for the trial and that speeches the previous week by Darrow and Malone had been "the greatest speeches of this century" (Scopes, 1925c). When the trial concluded with Scopes' conviction, Scopes made his first and only statement in the court transcript: "Your honor, I feel that I have been convicted of violating an unjust statute. I will continue in the future, as I have in the past, to oppose this law in any way I can. Any other action would be in violation of my ideal of academic freedom—that is, to teach the truth as guaranteed in our constitution, of personal and religious freedom. I think that the fine is unjust." Scopes was fined $100, plus $327.87 in court costs. Scopes did not pay the fine, but it remains unclear who did. According to court records, Scopes' $500 bond was guaranteed by Fidelity and Deposit Company of Maryland.

Scopes claimed that his trial was not meant "to decide any of [the issues in the headlines] ... but to bring a better understanding between the various factions involved" (Scopes, 1925a, b). He also believed that although his attorneys had "all the facts, logic, and justice," "these weapons are often ineffective in a battle against bigotry and prejudice" (Tompkins, 1965). He then vowed that although it had been "beyond my hopes" to win the case, "defeat will not be admitted unless the high court of the land goes against us.... Every possible interest will be fought out bitterly. Success is ultimately with us"

John Scopes being sentenced. On July 21, Scopes was found guilty of violating Tennessee's Butler Act and fined $100 plus court costs. Scopes' conviction was overturned 18 months later, but the Butler Act remained on the books until 1967. To the right of Scopes' head is Wallace Haggard, who is standing over Gordon McKenzie. Over Scopes' left shoulder is Bryan. The officer standing to Scopes' left is bailiff Kelso Rice, and Dudley Field Malone's head is in the lower right. Note WGN's microphone that was used to broadcast the trial (RCHGS).

(Scopes expected verdict of guilty, 1925). After declining an offer to continue teaching and coaching at RCHS (because he "did not want to be a cause of disturbance"), Scopes went to Lexington to see friends and investigate going to law school. He returned to Dayton when he heard that Bryan had died there, but he did not visit the grieving Bryan family there because, after seeing the large crowds gathered where Bryan had been staying, he decided, "The most Christian thing I could do was to keep out of the way and hope that Mrs. Bryan would learn from others of my sympathy for her and her family" (Tompkins, 1965). Scopes and his friends burned the bushels of mail that Scopes had received. Scopes later claimed that, except for his sister Lela being fired in Paducah for not renouncing his beliefs in evolution, his famous trial had "no other effect upon my family," but he was wrong (Moore, 2020).

On May 31–June 1, 1926, the Tennessee Supreme Court heard Scopes' appeal, but Scopes did not attend the hearing because he was busy in graduate school and had no interest in the case, adding that "it seemed to me that the trial at Dayton had accomplished as much as any legal restating could do.... I was as satisfied as the Tennessee justices to leave the case where it lay" (Scopes and Presley, 1967). During the appeal, the state invoked Bryan's majoritarianism, noting that "public schools are created by the legislature, and the courts can in no manner control, limit, or proscribe the legislature in the exercise of power over them.... The fact that a group of self-styled 'intellectuals' who call themselves 'scientists' believe that a certain theory or thing is true does not to any degree prevent the state legislature ... from forbidding the teaching or practicing of such a thing or theory which the legislature may conclude to be inimical ... to the general public welfare." Darrow responded by pleading for freedom and claiming, "We are once more fighting the old question, which after all is nothing but a question of the intellectual freedom of man."

On January 15, 1927, the Tennessee Supreme Court upheld the Butler Act as "a valid and constitutional law" but set aside Scopes' conviction because Judge Raulston, instead of the jury as required by law, had set Scopes' fine.[43] After claiming that the Butler Act "requires the teaching of nothing" and therefore doesn't give "preference to any religious establishment," the Court noted that Scopes "had no right or privilege to serve the state except upon such terms as the state prescribed." It then added that it could "see nothing to be gained by prolonging the life of this bizarre case. On the contrary, we think the peace and dignity of the State ... will be better conserved by the entry of *nolle prosequi* [i.e., not pursuing the case] herein. Such a course is suggested to the Attorney General." Tennessee Attorney General L.D. Smith immediately (and without comment) announced that Tennessee would not retry the case, thereby leaving no conviction for the defense to appeal. John Scopes' famous case was over. The ACLU searched for another defendant, but could not find one, and the Butler Act remained on the books for more than 40 years. Scopes was the only teacher ever prosecuted under the law.

After his trial, Scopes hoped that "my notoriety would ebb and I would be forgotten," but "those were false hopes.... Although I considered the [trial] and Dayton a closed chapter in my life, few others considered it so" (Scopes and Presley, 1967). Scopes refused to profit from his accidental fame on the lecture circuit because he feared that people would come to look at, instead of listen to, him. Scopes added, "Least of all did I want to make a quick dollar out of my experience; I had too much respect for the issues involved in the trial.... I knew that none of these offers ... could give me two things I wanted above all else: peace and emotional stability.... I could not live happily in a spotlight. The best thing to do, I was beginning to realize, was to change my life and seek anonymity" (Scopes and Presley, 1967). He would seek that anonymity in graduate school in another state, and then as a geologist on another continent.

After the Trial: John Scopes' Career as a Petroleum Geologist

Even before his trial began, Scopes suspected that his future in Dayton was over, admitting, "I am out of here and I have not the slightest idea what I shall do" (Many science teachers, 1925). Although some of his former professors at the University of Kentucky encouraged him to pursue law, Scopes feared a lifetime of comparisons with Darrow, and—using $2,550 raised by scientists at his trial—began graduate work in geology at the University of Chicago. There, he visited the Darrows often and later admitted that, except for his father, "Darrow had a greater influence on my life than any other man I have known" (Scopes and Presley, 1967). Scopes did not take any courses from the University of Chicago's experts who had come to Dayton to testify in his trial.

Scopes was an honor student but, after two years, he ran out of money. Longing to be "just another Yankee oil hunter" and "just another man instead of the Monkey Trial defendant," Scopes took a job as a field engineer with Gulf Oil in the jungles surrounding Lake Maracaibo in Venezuela. Scopes liked the work, but in 1928 he became seriously ill and returned to the United States. After recovering and joining his father for a nostalgic visit to England, Scopes resumed his geological work in Venezuela. There, in 1930, he met and married Mildred Elizabeth Walker. To please Mildred and unify the family (Mildred was Catholic), Scopes converted to Catholicism. However, for the rest of his life, Scopes remained an agnostic; he respected parts of religion, but never again committed himself to it (Shelton and Smith, 1979).

During his second stint in Venezuela, while working on the Orinoco River, Scopes was fired when he refused to go into Colombia, where Gulf Oil was not authorized to work and could not protect him. As Scopes later noted, "there was no prize that Gulf Oil could win that was worth the risk of going into a Colombian jail" (Silverman, 2018). Scopes returned to the United States and resumed his graduate work at the University of Chicago, studying under the direction of geologist Edson Sunderland Bastin (1878–1953). He spent much of 1931 doing field work in New Mexico, but soon thereafter financial problems again forced him to drop out of school. When he was told that his last hope for money, a scholarship to study the origin of oil and gas, was rejected and that he should "take your atheistic marbles and play elsewhere," Scopes again left college. He never earned a graduate degree (Scopes and Presley, 1967; Creviston, 2019).

By summer 1932, John and Mildred were living with John's parents in Paducah, where they welcomed their first son, John Thomas Scopes, Jr. Meanwhile, John was "a private in the army of the unemployed" (Central figure, 1932). Scopes and his parents continued to get letters about his trial for several years, and family members' homes were occasionally burglarized by thieves presumably hoping to find the money that newspapers had inaccurately claimed that Scopes had received from his famous trial. John, who by now considered his trial a "dead issue" that had accomplished nothing, told reporters, "I try to keep from thinking anything about it. I've lived with it so long I would like to forget about it" (Mack, 1966). While living in Paducah, John ran for Congress as a Socialist, but he was soundly defeated (Scopes of evolution fame seeks Congress seat, 1932).

In the Summer of 1933, Scopes began working for United Production (later known as United Gas) Corporation as an oil scout in Beeville, Texas, and then in Houston, where his son Bill was born in 1936. In 1940, Scopes and his wife moved to United's corporate offices in Shreveport, Louisiana, where they lived for the rest of their lives. Scopes, who described himself as "Jack-of-all-trades and a specialist in none," worked on oilfield economics, regulatory compliance, taxes, gas reserves, pipeline expansion, appraisals, and pipeline expansion. He testified in Washington, D.C., and elsewhere about petroleum reserves. Throughout his life, Scopes enjoyed what he called "the normal, ordinary work characteristic of any large oil and gas company. There were no highlights and I had the same number and the same kind of experiences as anyone else who did that type of work." According to Scopes, "not once did anyone connect geologist John T. Scopes with the man in the 'monkey trial'" (Doyle, 1960). Scopes, whose only college degree was in law, had a productive career, but never earned a geology degree and never joined the American Association of Petroleum Geologists (Silverman, 2018). In 1955, the Tennessee Historical Commission unveiled an error-filled historical marker in front of the courthouse commemorating Scopes' trial, but Scopes did not attend the ceremony (Moore, 2021).

After dropping out of graduate school at the University of Chicago, John began working for Gulf Oil in the jungles of Venezuela (Clarence Darrow Collection, Rienenfeld Rare Books Research Center, Law Library, University of Minnesota).

Over the years, Scopes did not run from the press; his name, address, and phone number were listed in the telephone directory, and his name was prominent on his roadside mailbox. However, Scopes seldom talked about his trial, even with his family, and he rarely gave interviews, for he had realized, "Every time someone makes an effort to meet me, I know it is a holy man" (de Camp, 1968). Scopes understood that he could neither escape nor voluntarily relinquish his fame and notoriety, but he remained politely uncooperative with most reporters. As Scopes' friend Jerry Tompkins noted, "No public figure who actually achieved a second anonymity ever welcomed it more than John Scopes" (Tompkins, 1972). Scopes only occasionally commented on his case:

> 1951: Scopes condemned the Butler Act as "a stupid law when I was tried for it … [and] it is a stupid law today.… [Tennessee] must repeal it before Tennesseans can have any respect for themselves" (Wesolowski, 1975).
>
> 1960: Scopes told author and broadcaster Studs Terkel that "I'm sure I did [teach evolution] if I even substituted in a class in biology for a period of three weeks." Terkel then asked Scopes if he liked the way that Dick York played him in *Inherit the* Wind, to which Scopes quickly replied, "he didn't play me." When asked about his view on progress since his trial, Scopes admitted that "we haven't advanced too much.… No, I don't think we have" (John T. Scopes discusses, 1960).
>
> 1963: Scopes wondered "why in this space age and in view of the advancements science has achieved [are we] arguing over something archaic like [the Butler Act].… The Bible is a book of morals, not a textbook. It shouldn't be used as a yardstick of science." Although in 1925 Scopes had believed that he had won his case, he admitted in 1963 that "the older I get the more I realize it was no victory" (Scopes advises two teachers, 1963).
>
> 1965: Scopes told a Canadian television host that he "didn't care if people in Tennessee believed in evolution or didn't believe in evolution. The question wasn't whether evolution or the Bible was right. The issue was whether a person's liberties should be infringed upon by any society" (Front Page Challenge, 1965).
>
> 1965: When Arkansas Governor Orval Faubus (1910–1994) praised his state's ban on the teaching of human evolution and proclaimed that the Bible is "good enough for me," Scopes warned that "no politician is going to go out on a limb and try to repeal" the Arkansas law. Scopes urged Arkansans to "figure out what the governor is thinking so you can protect [yourself] from the governor.… You don't protect any of your individual liberties by lying down and going to sleep" (World famous teacher, 1965).
>
> 1968: When the U.S. Supreme Court ruled in *Epperson v. Arkansas* that laws banning the teaching of human evolution in public schools are unconstitutional, Scopes told reporters that "this is what I've been working for along.… I'm happy about the decision. I thought all along—ever since 1925—that the law was unconstitutional" (A Darwinism ban in Arkansas ends, 1968). When they met in January 1969, Scopes confided to Susan Epperson that he was still receiving an average of one letter per day about his 43-year-old trial. Epperson had received dozens of letters about her trial, but Scopes got even more letters about her case than did Epperson.

Inherit the Wind and Retirement

When the play *Inherit the Wind* premiered on Broadway on April 21, 1955, Scopes was offered tickets, but he declined to attend because he was "too busy and not interested enough to fly to New York to see it." However, when the movie version had its North American premiere in Dayton on July 21, 1960, at Scopes Trial Day, Scopes—at the urging of his wife—was there. Scopes "enjoyed the movie" and admitted that it "captured the emotions in the battle of words between Bryan and Darrow" (Scopes and Presley, 1967). At the day-long celebration of the 35th anniversary of Scopes' trial, Scopes was honored by a parade and reunited with many of his old friends, including

From 1925 until he died, John Scopes received thousands of requests for his autograph. Although he ignored most requests because they weren't "even courteous enough to enclose a stamp," the request shown here was successful. This is John Scopes' autograph on an envelope sold on the first day of issue (July 1, 1957) of a 3¢ U.S. postage stamp honoring America's teachers (Randy Moore).

prosecutors Sue Hicks and Wallace Haggard. While in Dayton, Scopes visited Robinson's Drug Store, where he enjoyed a "Scopes Soda" priced 15¢ "Now as Then in Honor of Scopes' Return to Dayton." Scopes admitted that he "would do the whole thing over again," and juror James Bowman claimed that Scopes would again be convicted if the trial were repeated. Scopes agreed (Scopes wins acclaim, 1960).

By 1964, Scopes was tired and "burned out," and he decided to retire. After writing a well-received article for Jerry Tompkins's *D-Days at Dayton: Reflections on the Scopes Trial* (1965), Scopes began working with James Presley, who produced *Center of the Storm: Memoirs of John T. Scopes* (1967). Scopes made several public appearances promoting the book, adding, "If nothing else, it was time to remind the people again of the importance of preserving their basic freedoms" (Olasky and Perry, 2005). In his later years, Scopes continued to defend his convictions:

> Teaching science: "If we were still trying to teach science based on the Bible, we'd still be driving a horse and buggy" (John T. Scopes discusses, 1960).
>
> Teachers' rights: "Especially repulsive are laws restricting the constitutional freedom of teachers. The mere presence of such a law is a club held over the heads of the timid.... By limiting freedom, [these laws] are helping to make robot factories out of schools" (Scopes and Presley, 1967).
>
> Personal liberty: Scopes defended the rights of fundamentalists to their beliefs, but rejected their insistence that others also hold those beliefs. Throughout his life, Scopes viewed his trial as an example of the ongoing conflict between intolerance and broad-minded tolerance which, he added, could "be seen in all lands, in all ages."
>
> Church, state, and schools: "It is as bad to have school and state joined as it is to have church and state joined" (John Scopes says, 1965).
>
> Higher education: Scopes believed that "there is more intolerance in higher education than

in all the mountains of Tennessee" (Scopes and Presley, 1967) and urged college students to "demand teachers with some sense. What they get in the classroom is researchers, scientists. A research scientist is the poorest goddam teacher in the world. My god, let's get some teachers who can teach" (Scopes and Presley, 1967; Shaw, 1970).

His religious faith: Although he had declared himself a Christian just before the start of his trial, Scopes later claimed that anyone who knew that they were a Christian "probably had a different view of what it takes to be a Christian than I did. It has been my observation that anyone who is sure he is a Christian is likely to be mistaken" (Tompkins, 1965).

Organized religion: Scopes believed that fundamentalists had "polluted" Christ's teachings and that "churches as a whole—if they do any good, they do so much harm that they more than counterbalance their value" (Front page challenge, 1965).

Scopes also became concerned about pollution: "All this talk about reclaiming the environment through legislation and technology know-how: It won't work. Pollution control is a matter of getting at the cause, not the effect. The cause is human. It's people we've got to change. We have learned to worship at the altar of technology—and it's going to destroy us" (Tompkins, 1972).

John Scopes' Final Days

In April 1970, Scopes visited Nashville's George Peabody College for Teachers, where he made his first appearance in a Tennessee biology classroom in 45 years.[44] While there, Scopes—still admitting to be an "old-time Socialist"—revisited his trial's important issues, including individual rights, academic freedom, and what he considered to be his small part in the trial. He announced that he supported sex education and prayer in classrooms, and warned that the United States was "controlled to a large degree by fanatics" (Scopes returns, 1970). Scopes' day included a talk with biology students, a luncheon, a press conference, and an afternoon lecture. Scopes talked with politicians, reporters, friends, and admirers as he discussed his trial and current affairs. He was greeted by his college roommate (John Stark, also a retired geologist) and Rhea County judge June Forman, and was given a certified copy of the bill that repealed the Butler Act. When reminded that he had stayed in Dayton after his contract expired in 1925 to pursue a blond who deserved "as much credit as anyone for the trial being held in Dayton that summer," Scopes admitted that he could no longer recall her name (Scopes and Presley, 1967). Scopes also addressed more serious issues. After warning reporters that the country was "moving toward fascism," Scopes added that he had "voted for that fellow [George] Wallace. He's no more racist than the rest, and he might scare somebody into doing something.... I think most of these civil-rights agitators, the whites, I mean, are motivated by a guilty conscience." Scopes then surprised many in the crowd by pleading for allowing prayers in classrooms, adding, "If I were teaching school and the only way I could communicate with the class was through prayer, I would pray. And it would be none of the Supreme Court's business" (Shaw, 1970).

Scopes' talk in Nashville was his last public appearance. Three months later, he was diagnosed with inoperable esophageal cancer. His last words were to his wife Mildred: "I love you."

Scopes died at age 70 of esophageal cancer at his home in Shreveport, Louisiana, on October 21, 1970. His death was news the world, including on the front page of the *New York Times* (Scopes of "Monkey Trial" is dead at 70). Following services in

On April 1, 1970, John Scopes made his first appearance in a Tennessee classroom in more than four decades when he spoke at the George Peabody College for teachers in Nashville. This was Scopes' last public appearance (Robert Johnson, USA Today Network).

Shreveport and Paducah, Scopes was buried in a family plot in Oak Grove Cemetery in Paducah. His sister Lela, who is buried beside him, provided his epitaph: "A Man of Courage" (Mr. John Scopes, 1970; Craig, 1979). Just outside the cemetery, a historical marker commemorates Scopes' grave.

Three generations of Scopeses. Left to right: Deborah Scopes (Bill and Jackie Scopes' daughter), Mildred Scopes (John's wife, holding Deborah), Bill Scopes, and John Scopes in Shreveport, Louisiana, in August 1970. This is the last photograph of John Scopes (Richard and Pat Heflin).

Sources

Anderson, Paul Y. 1925. The drugstore argument that swelled into a national debate. *St. Louis Post-Dispatch*, May 31, 17.
Bailey, Kenneth Kyle. 1953. The Anti-Evolution Crusade of the Nineteen-Twenties. Ph.D. Dissertation, Vanderbilt University, Nashville, Tennessee, 125.
Baptist editor supports Bryan. 1925. *Nashville Banner*, June 15, 1.
"Battle of Dayton" on again, says Neal. 1927. *New York Times*, September 12, 14.
Central figure in famous Scopes case jobless now. 1932. *The News-Palladium* (Benton Harbor, MI), October 4, 11.
Craig, Berry. 1979. Scopes revisited. *Paducah Sun*, August 10, 1, 6.
Creviston, J. 2019. E-mail to Randy Moore, dated November 12, 2019, from Office of the University Registrar, University of Chicago.
Darwin in Scopes' place for a while. 1925. *Chattanooga News*, August 18, 1.
A Darwinism ban in Arkansas ends. 1968. *New York Times*, November 13, 1, 4.
Dayton, Tenn., put "on map" by evolution trial. 1925. *Des Moines Register*, May 26, 6.
de Camp, L. Sprague. 1968. *The Great Monkey Trial*. New York: Doubleday, 466.
Doyle, Jim. 1960. John Scopes would teach it again. *Ft. Worth Star-Telegram*, October 21, 6.
Evolution trial opens this morning; Raulston relies on divine guidance. 1925. *Chattanooga Daily Times*, July 10, 1.
Football, evolution, and bootlegging in east Tennessee.1925. *Sioux City (Iowa) Journal*, June 8, 4.
Front Page Challenge. 1965. John Scopes, "Monkey Trial" defendant. Canadian Broadcasting Company, October 5. Accessed November 16, 2020, at https://www.cbc.ca/archives/entry/darwin-john-scopes-monkey-trial-defendant, 8:59, 9:10.
Garrett, Oliver H.P. 1925. Scopes seeks information on evolution. 1925. *Chattanooga Times*, June 13, 1.
Ginger, Ray. 1958. *Six Days or Forever? Tennessee v. John Thomas Scopes*. London: Oxford University Press, 101.
John Scopes says "I'd do it again." 1965. *Waterloo (Iowa) Courier*, November 14, 4.
John T. Scopes discusses the "Scope Monkey Trial" of 1925. 1960. Studs Terkel Radio Archives, October 20, 14:07.
Lait, Jack. 1925. Lait asserts conviction is certain fate. *Minneapolis Daily Star*, July 11, 1.
Mack, Darrell. 1966. "Monkey Trial" fizzles on lack of sensation. *Shreveport Times*, April 2, 9A.
Many science teachers seek Scopes' position. 1925. *Chattanooga Daily Times*, July 19, 1.
Milton, George F. 1925. Scopes likes Tennessee town better than New York. *Baltimore Evening Sun*, June 16, 2.
Mr. John Scopes, man of courage. 1970. *Nashville Tennessean*, November 23, 12.
Monkey war stirs Tennessee. 1925. *Owensboro (Kentucky) Inquirer*, May 17, 7.
Montgomery, Dennis. 1973. There's no John Scopes memorial in Salem. *Decatur (IL) Herald*, May 1, 15.
Moore, Randy. 2020. The long and lingering shadow of the Scopes Trial. *American Biology Teacher*, 82 (2), 81–84.
______. 2021. The most famous historical marker in Rhea County. *Rhea Notes*, 23 (3), 1–3.
Moutoux, John T. 1925. Accused evolution prof most popular man in town. *Knoxville News*, May 11, 1.
Of Monkeys and Men: Public and Private Views of the Scopes Trial. 2021. Special Collections, University of Tennessee, Knoxville, Tennessee.
Ogden, Warner. 1960. Scopes challenges wording of marker. *Knoxville News-Sentinel*, July 24, A8.
Olasky, Marvin and John Perry. 2005. *Monkey Business*. Nashville: Broadman & Holman, 14, 175.
Other teachers equally guilty as Scopes-Neal. 1925. *Chattanooga News*, May 18, 1.
Potter, Charles Francis. 1951. *Preacher and I*. New York: Crown, 269.
Rhea High Fundamentalists have successful campaign. 1925. *Chattanooga Times*, November 29, 25.
Rhea High gridders open season Friday. 1922. *Chattanooga Daily Times*, September 15, 8.
Rhea's green team makes big strides. 1923. *Chattanooga Daily Times*, December 9, 19.
Scopes advises two teachers to go to court. 1963. *Chattanooga Times*, April 5, 1, 7.
Scopes applies for job. 1925. *Nashville Banner*, June 17, 1.
Scopes, awaiting trial, is keeping cool. 1925. *Walla Walla Daily Bulletin*, July 1, 7.
Scopes, back home to see his mother, says Darrow can outwit prosecution. 1925. *Paducah Evening Sun*, July 3, 1, 9.
Scopes ill at ease when stammering brief explanation. 1925. *Kingsport (Tennessee) Times*, June 11, 1.
Scopes, John T. 1925a. Letter to the editor. *Forum*, 74 (2), 3.
______. 1925b. Scopes expected verdict of guilty. *Minneapolis Daily* Star, July 22, 8.
______. 1925c. Scopes writes own story of trial. *San Francisco Examiner*, July 19, 2.
______, and James Presley. 1967. *Center of the Storm: Memoirs of John T. Scopes*. New York: Holt, Rinehart and Winston, 18, 41, 70, 109, 134–136, 188, 194, 208, 220, 233–235, 239–243, 266–276.
Scopes of evolution fame seeks Congress seat. 1932. *Chicago Tribune*, August 13.

Scopes of "Monkey Trial" is dead at 70. 1970. *New York Times*, October 23, 1.
Scopes returns. 1970. *Time*, 95 (15), 5.
Scopes tells of religious beliefs. 1925. *Nashville Banner*, June 17, 1.
Scopes wins acclaim on return to Dayton. 1960. *Knoxville Journal*, July 22, 5.
Shannon, Elaine. 1970. Scopes asks for fight to continue. *Nashville Tennessean*, April 2, 1, 14.
Shaw, Bynum. 1970. Scopes reviews the Monkey Trial. *Esquire*, 74 (Nov.), 86–89.
Shelton, Sharon and Mary Ann Smith. 1979. Lela Scopes. McCracken County (Kentucky) Public Library, Digital Collections. Oral Histories-KOHC 1974–1979; Filename: 1976OH79.21-Lela Scopes.mp3, 7:58, 25:35–49.
Silverman, Matt R. 2018. John T. Scopes: A summer in hell and a career in petroleum geology. *AAPG Explorer*, 39 (4), 20–21.
Snook, Sidney. 1925. Scopes here to rest, in "Monkey Fight" to the end he declares. *Paducah Sun-Democrat*, May 16, 1.
Snyder, Julian. 1925. Scopes tires of publicity. *Nashville Banner*, June 7, 1.
Stone, Irving. 1941. *Clarence Darrow for the Defense*. Garden City, NY: Doubleday, 494.
Strongen, Ted. 1960. Never taught evolution in Dayton, Scopes says. *Chattanooga Daily Times*, July 22, 1, 2.
Taking his conviction lightly. 1925. *Lexington Herald*, May 26, 1.
Tompkins, Jerry R. 1965. *D-Days at Dayton: Reflections on the Scopes Trial*. Baton Rouge: Louisiana State University Press, 17–31.
______. 1972. Memoir of a belated hero. *American Biology Teacher*, 34 (7), 383–384, 410.
Wesolowski, James Walter. 1975. Before Canon 35: WGN broadcasts the Monkey Trial. *Tennessee Historical Quarterly* 34 (4), 392–406.
World famous teacher of evolution says repeal of state ban impossible. 1965. *Arkansas Gazette*, September 22, 18B.

Scopes, Lela V
1896–1989

Lela V Scopes, John's youngest sister, was born on October 30, 1896, in Paducah, Kentucky. Lela was named after her mother's younger sister, Lela Dora Brown Ruoff (1877–1939). After attending elementary school in Arcadia, Kentucky, Lela was moved to a private school in Lone Oak closer to her home. She then moved with her parents and brother John to Danville, where she graduated in 1916. Lela attended Murray State University, Western Kentucky University, and the University of Kentucky, where she was a student during her brother's trial. Because she and John attended college at the same time, they decided to reduce the financial strain on their parents by having Lela work as a teacher while John attended college full-time. (After John graduated, Lela enrolled full-time.) Lela, who at the University of Kentucky was a member of Kappa Delta Pi (an international honor society in education), never completed a master's degree, probably because she, like brother John, enrolled in courses based on the professor rather than a specific set of graduation requirements. In Fall 1922, Lela began teaching math at Paducah's Augusta Tilghman High School, where her work was "very satisfactory." Lela helped file the job applications that led to John's hiring in 1924 in Dayton.

On July 1, 1925, Lela was summoned to the office of Superintendent Lyda J. Hanifan, who asked Lela to refute her brother's claims about evolution. When Lela refused, she was fired. Hanifan explained: "I am opposed to the teaching of evolution in the public schools, and will not permit any teacher on our staff to give their pupils such instruction.... I think it is dangerous to teach the doctrine of evolution to pupils whose minds are not matured.... I will not have evolution taught in our schools under any consideration" (Federation asks facts, 1925). Schoolboard member Z.A. Bennett added that it "might be embarrassing to [Paducah's] public school system to become involved in the publicity resulting from the Scopes discussion" (Scopes' sister loses school position,

1925). Although many newspapers editorialized that a "very grave and unjustified injustice has been done Miss Scopes," Paducah's First Baptist Church supported Hanifan and condemned newspapers' attacks on Scopes' firing (The Lela Scopes case, 1925; Teacher barred by beliefs, 1925; Paducah Baptists rap evolution, 1925).

Lela's firing was a national news story. Lela, who believed that the Biblical story of creation complemented the "scientific facts of God's creation," never complained publicly about her firing, but she did wonder what her views of evolution had to do with teaching math. Lela's brother John, meanwhile, had regrets, telling reporters, "If I had known that this thing would make it hard for her to get a position, I would never have become involved in the case" (Scopes back home, 1925). In Louisville, Kentucky, rumors that a local school would soon hire Lela elicited protests by local preachers (Offer to Miss Scopes, 1925).

In 1925, Lela V Scopes—John Scopes' sister—lost her job as a math teacher in Paducah, Kentucky, when she would not renounce her brother's claims about evolution. This photograph shows Lela in college (Jeanette Gilliam Travis).

Lela soon had several job-offers. She accepted one at Highland Manor, a private school for girls in Tarrytown, New York, where superintendent Eugene Lehman told reporters that he "was impressed with the heroism of [Lela's] position and decided I should like to have her on our teaching staff." Lela taught math and biology, was paid $1,000 per year plus all expenses, and was promised opportunities to become registrar or dean of the school (Lela Scopes to teach in Tarrytown school, 1925). After two years in New York, Lela taught math in Winnetka, Illinois, where she got most of her on-the-job training and, according to the 1940 U.S. Census, lived with her partner, Edith McKay. During some of her years in Winnetka, John and Mildred Scopes' two teenaged sons lived with Lela, an experience that John Scopes, Jr., later claimed "saved us" (Moore, 2020). Lela helped pay for the boys' college educations.

Lela loved to travel with her family and friends. In 1937, she spent her summer break in England, meeting her father's family and seeing the area where he grew up. She returned to the United States aboard *RMS Berengaria* on September 3, 1937. She later admitted, "My name Scopes has brought pleasure in life rather than deprivation.

Only intelligent people recognized the name. I've made many interesting and lasting friendships as a result" (Scopes and Presley, 1967).

Lela retired in 1956 and rejoined her three sisters in Paducah, where she was active in a variety of civic groups. She died at her home in Paducah on January 23, 1989, and was buried beside her famous brother in Paducah's Oak Grove Cemetery beneath the epitaph, "A Gracious and Generous Lady."

Sources

Federation asks facts in Paducah evolution case. 1925. *Paducah Evening Sun*, July 3, 1, 9.
The Lela Scopes case. 1925. *Paducah News-Democrat*, July 4, 4.
Lela Scopes to teach at Tarrytown school. 1925. *Middletown Daily Herald*, August 15, 2.
Moore, Randy. 2020. The long and lingering shadow of the Scopes Trial. *The American Biology Teacher*, 82 (2), 81–84.
Offer to Miss Scopes draws ministers' fire. 1925. *Louisville Courier-Journal*, July 14, 1.
Paducah Baptists rap evolution. 1925. *Lexington Herald*, July 14, 1.
Scopes, back home to see his mother, says Darrow can outwit prosecution. 1925. *Paducah Evening Sun*, July 3, 1, 9.
Scopes, John T., and James Presley. 1967. *Center of the Storm: Memoirs of John T. Scopes*. New York: Holt Rinehart and Winston, 234.
Scopes' sister loses school position on evolution belief. 1925. *Miami News*, July 2, 8.
Teacher barred by beliefs, claim. 1925. *Louisville Courier-Journal*, July 2, 3.

Scopes, Thomas | Mary Alva
1860–1945 1865–1957

Thomas Scopes was born on June 23, 1860, in Middlesex County on the outskirts of London, England. He was the seventh of eight children to William Scopes (1818–1866) and Mary Turner Lucas Scopes (1822–1873). After his parents died, Thomas lived with his uncle in Needham Market in Suffolk, England. Thomas attended an Episcopalian school, sang in the choir at a nearby Catholic church, and was a voracious reader. Thomas, who served an apprenticeship in England as a marine machinist, was raised in an educated, working-class family (Shelton and Smith, 1979).

On February 28, 1882, Thomas left England for the United States, arriving in Philadelphia, Pennsylvania, with only some money, clothes, personal effects, and a few books (including a hymnal, a family Bible, a book about the French Revolution, and Charles Darwin's *On the Origin of Species*). Thomas travelled across the United States and, while in Texas, saw an ad in a newspaper that convinced him to buy shares in the Shelton Machine Shop in Union City, Tennessee. However, by the time he got to Tennessee, the shop's owner had died, and he gave his shares to the widow (Shelton and Smith, 1979).

By 1883, Thomas was working in Paducah as a machinist for the Illinois Central Railroad. According to one family story, Thomas made a bell for legendary railroader Casey Jones, an engineer for the Illinois Central Railroad.[45] Thomas planned to work for a while, explore the United States, and return home, but—except for a short, nostalgic return to England in 1929 to introduce John to his family there—he stayed in the United States for the rest of his life. Thomas became a naturalized citizen of the United States in 1899.

Mary Alva Brown, the daughter of a Presbyterian preacher, was born in Fulton, Kentucky, on July 15, 1865, to John Thompkins Brown (1829–1898) and Martha Elizabeth

Calvert Brown (1838–1920). Although Mary Alva was taken out of school early because of vision problems, she subsequently taught herself to read and write. Mary Alva married Thomas Scopes on October 15, 1885, and soon thereafter they started a family that included six children (Appendix 1):

Charles Calvert Scopes died before reaching his first birthday. Charles is the only one of Thomas and Mary Alva Scopes' children who is not buried in Paducah.

Mary Irene Scopes was a member of the first graduating class of Western Kentucky University (then named Western Kentucky Normal School) in Bowling Green. She then taught elementary school (for $60 per month) and was active in various social clubs. In 1914, Irene married Edgar Metzger, whose family owned the Metzger Meat Packing Plant.

Ethel Elizabeth Scopes also taught elementary school. She married Thomas T. Clark on Christmas Day, 1913, and by 1930 the couple and their children were living next door to Thomas and Mary Alva on Berger Road in Paducah. Ethel, Irene, and Lela were important in the lives of John and Mildred Scopes' two sons, who lived with them while their parents were having marital problems (Moore, 2020).

Nannie Mae Scopes graduated from high school in 1906 and married Walter Gilliam on Christmas Day the next year. She was active in Paducah's civic groups and taught math as a substitute teacher. In 1927, with older brother Cleve as the primary owner, Walter and his three brothers opened the Gilliam Candy Kitchen (Clark, 1937).

Lela V Scopes became a math teacher and lost her job in Paducah in 1925 when she would not renounce her brother John's views about evolution. She is discussed in a separate entry in this book.

John Thomas Scopes became the namesake for the famous "Monkey Trial" in Dayton, Tennessee, in 1925. He is discussed throughout this book.

According to John, Thomas and Mary Alva were "a good team" in which Mary Alva was the "business head and balance wheel" (Scopes and Presley, 1967). Except for Charles,

Thomas and Mary Alva Scopes' family, in late 1900. Left to right: Mary Irene, Nannie Mae, Thomas, Lela V (in lap), Ethel Elizabeth, John Thomas (baby in lap), and Mary Alva (Jeanette Gilliam Travis).

all of the Scopes' children took piano and elocution classes and graduated from high school. Four of the children earned college degrees, with both Lela and John completing some post-graduate work.

Thomas and Mary Alva were initially active in the Cumberland Presbyterian church, but Thomas—a deacon—broke with the church when it tried to rid the area of prostitutes. Thomas, "a dissenter by habit," wanted to help the women rather than run them out of town. Thereafter, Thomas was an agnostic who had a "Christian approach to life" (Scopes and Presley, 1967). The Scopes' home was intellectually vibrant; according to their daughter Ethel, Thomas "liked to talk, he dwelled on science, social economy and all subjects that stand for the betterment of humanity. There are few people who [could] see the possibility of a 'Kingdom on Earth' as confidently as my father could" (Clark, 1937).

In England, Thomas had not been part of the labor movement, but the 1894 Pullman Strike in the United States "nudged Dad into the ranks of organized labor" (Scopes and Presley, 1967). Thomas became involved with the International Association of Machinists and the American Railway Union, and was fired from his job as a roundhouse foreman when he refused to be a strike breaker. Although he was later rehired, Thomas's support of the Pullman workers demonstrated his acceptance of personal sacrifice for what he considered to be public good. He instilled this conviction in his children; in 1925, his son John would take a stand for academic freedom in Dayton.

Thomas's activism was also reflected in his running for political offices—for example, for city judge in 1901, and for Presidential elector in 1904—on the Socialist Labor Party ticket (Socialist nominations, 1901). Thomas often interacted with Eugene Debs, a socialist who founded the American Railway Union. During the Pullman Strike, Chicago attorney Clarence Darrow—who would defend Thomas's son John in Dayton decades later—defended Debs and other union officials (Darrow, 1932). In 1910, when Debs spoke to nearly 3,000 people in Paducah, he was introduced by Thomas Scopes (Debs' meeting, 1910). In 1932, following in his father's footsteps, John Scopes ran for Congress as a Socialist. As John later noted, "in Kentucky, our family's way of thinking … [wasn't] a generally accepted one" (Scopes and Presley, 1967).

From 1913 to 1916, Thomas and his family lived in Danville, Illinois, where Thomas worked in the Brewer Railway Yard. Thomas, Mary Alva, and their two youngest children—Lela V and John (the others had established lives of their own)—then moved to Salem, Illinois. There, they met and often heard speeches by hometown hero William Jennings Bryan. In 1919, when John graduated from Salem High School, the commencement speaker was Bryan. Six years later, in Dayton, Thomas and John would again see Bryan, who helped the state prosecute John.

In July 1925, Thomas went to Dayton to support his son during his trial. While there, Thomas boasted, "I'm for my boy, and I'm for evolution, too.... I'm for him, not only as a dad, but as an evolutionist. This boy hasn't done anything to be tried for. He's a chip o' the old block and I'm proud of him." He later told Bryan, "I endorsed every word of what you said at my son's [high school] commencement exercise, but I can't agree with you now." When Bryan replied, "Perhaps you're biased," Thomas responded, "Yes, perhaps we both are" (Ogden, 1960).

Thomas stressed to reporters that he, like his son, was an evolutionist because "nobody who reads a lick can be anything else but an evolutionist, and I claim I'm as good a Christian as the rest of 'em, too." He also admitted that his trip to Dayton was his

Soon after arriving in Dayton for his son's trial, John Scopes' father Thomas (right) told reporters, "No body who reads a lick can be anything else but an evolutionist," adding, "I'm as good a Christian as the rest of 'em, too." In this photograph, Thomas poses with his son in Dayton (RCHGS).

first vacation in 20 years, and that it was "kinda funny, taking a vacation to see your son go on trial as a criminal" (Thomas Scopes stands by his son, 1925). John later claimed that his father "viewed the struggle as I did. He knew that freedom was at stake … and he considered the trial to be just as much his as mine.… He was happy that I was part of an event that he felt was a service to the country" (Scopes and Presley, 1967).

Thomas later went blind and became less active in politics and organized labor. He died in Paducah on May 27, 1945, after which Mary Alva moved in with her daughter Ethel in Paducah. Mary Alva, who often boasted about her 12 grandchildren and 13 great-grandchildren, died on October 1, 1957, and was buried beside her husband in Paducah's Oak Grove Cemetery.

Sources

Clark, Ethel Scopes. Ca. 1937. *Room for One More*. Unpublished manuscript, owned by Jeanette Gilliam Travis, Paducah, Kentucky.
Darrow, Clarence. 1932. *The Story of My Life*. New York: Charles Scribner's Sons, 58.
Debs' meeting: Address on Socialism at auditorium—Mr. Scopes presides. 1910. *Paducah Sun*, May 24, 3.
Moore, Randy. 2020. The long and lingering shadow of the Scopes Trial. *The American Biology Teacher*, 82 (2), 81–84.
Ogden, Warner. 1960. Scopes Trial figures recall old memories. *Knoxville News-Sentinel*, July 17, B10.
Scopes, John T., and James Presley. 1967. *Center of the Storm: Memoirs of John T. Scopes*. New York: Holt Rinehart and Winston, 5–12, 19, 22–25, 68.
Shelton, Sharon, and Mary Ann Smith. 1979. Lela Scopes. McCracken County (Kentucky) Public Library, Digital Collections. Oral Histories-KOHC 1974–1979; Filename: 1976OH79.21-Lela Scopes.mp3; Identifier: 1976OH79.21.
Socialist nominations 1901. *Paducah Sun*, May 28, 4.
Thomas Scopes stands by his son. 1925. *Paducah News-Democrat*, July 12, 1.

Scopes Trial Day
July 21, 1960

On Thursday, July 21, 1960, Dayton celebrated the 35th anniversary of the Scopes Trial by hosting Scopes Trial Day. Throughout town, merchants decorated their stores as if it were 1925. Attendees included John and Mildred Scopes; Watson Davis of Science Service, who covered the trial and helped recruit the scientists that Darrow tried to have testify; Helen Rudd Brown, William Jennings Bryan's granddaughter and vice-chair of the California State Democratic Committee; Jesse Goodrich, W.G. Taylor, and J.H. Bowman, jurors at the trial; Vera Raulston, the daughter of Judge John Raulston; Sue Hicks and Wallace Haggard, who prosecuted Scopes; Adolph Ochs, Earl Shaub, Robert Kyle, Nellie Kenyon, Warner Ogden, and Emma McSmith, reporters who covered the trial; J.J. Rodgers, mayor of Dayton; Theodore Mercer, president of Bryan College; and Jason Rudd, president emeritus of Bryan College.

The day's events included a press conference for Scopes and other trial figures, a parade, presentation of a Key to the City to Scopes by Rodgers, tours of the courtroom, a band concert, fashion parade, and a piano recital. In town, Robinson's Drug Store again sold root beer for 5¢ and "Scopes Sodas" for 15¢, "Priced Now as Then in Honor of Scopes's Return to Dayton." The event culminated with the first U.S. showing of *Inherit the Wind* (Scopes Trial Day, 1960). Before the movie started, Scopes told the crowd that he "had hoped the trial would accomplish more than it did." At the drive-in, the festivities ended with a town meeting to discuss the movie.

On July 21, 1960, the 35th anniversary of John Scopes' conviction, Dayton celebrated Scopes Trial Day. The festivities included a parade, a speech by John Scopes, the American premiere of *Inherit the Wind*, and a reunion of Scopes, Wallace Haggard, Sue Hicks, some jurors, and journalists who had covered the trial (RCHGS).

When giving Scopes the Key to the City, Mayor Rodgers told the crowd, "Symbolically, as well as actually, all hard feelings, rancor, and lingering passions which arose from the famous Scopes case will be buried, and the way paved for Dayton to go onto a brighter tomorrow, as it smilingly recalls the notoriety of long ago" (Scopes Trial Day, 1960; Figs. 44, 45). Scopes' wife, Mildred, admitted that she "had often wondered how the people of Dayton felt about my husband, and if there were any hard feelings because of the flood of publicity which he inadvertently brought to the town when he agreed to test the state law concerning evolution." Mildred added that she was "astonished that [John] came. He never talks about it.... He is very shy—very publicity shy. This [event] has been good—it was a very kind and gracious thing for the town to do and we appreciate it so much" (Hardaway, 1960).

After the showing of *Inherit the Wind*, a reception for director Stanley Kramer was held at the home of Adele and George Shibley, who owned a department store in downtown Dayton. At the reception was Dayton resident Eloise Reed, who had attended the Scopes Trial as a 13-year-old girl. Reed later recalled that when Kramer asked her if she liked the movie, she replied, "I didn't like it. I thought it was awful ... [Dayton] wasn't like that at all" (American Experience, 2002).[46]

During Scopes Trial Day, John Scopes (right) was given a key to the city by Dayton mayor J.J. Rodgers, who "smilingly recall[ed] the notoriety of long ago." Standing between mayor Rodgers and John Scopes is Scopes' wife, Mildred (RCHGS).

While in Dayton, Scopes addressed a large crowd from atop the steps of Rhea County Courthouse, where he had been convicted of teaching human evolution 35 years earlier (RCHGS).

John and Mildred Scopes, anxious to see their grandchildren, left for New Orleans soon after the events concluded. Scopes Trial Day produced the second-largest crowd in Dayton's history.

Sources

American Experience. 2002. *Monkey Trial*. Boston: WGBH Educational Foundation, 1:16:55–1:17:18.
Hardaway, Sybil. 1960. Dayton key important trophy to Mrs. Scopes. *Chattanooga News/Free Press*, July 21.
Scopes Trial Day in Dayton Thursday amid big celebration. 1960. *Dayton Herald*, July 21, 1.

Scopes Trial Grove
Est. 1925

The Scopes Trial Grove is a group of maple (*Acer nigrum*) and willow oak (*Quercus phellos*) trees on the Rhea County Courthouse lawn that were saplings at the time of the Scopes Trial. The trees, which are recognized as historic trees by the Tennessee Urban Forestry Council, are among the largest of their species in Tennessee. Some of the trees are marked with plaques stating, "This tree witnessed the 1925 Scopes Trial."

Source

Scopes Trial Grove #1. 2018. Tennessee Urban Forestry Council, January 17. Accessed November 3, 2020, at https://www.tufc.com/project/scopes-trial-grove/.

Scopes Trial Play and Festival
Est. 1988

In 1988, Frank Chapin—a drama director from Chattanooga—visited Dayton with plans to produce *Inherit the Wind* in the Rhea County Courthouse. He abandoned the idea when he learned of Dayton's dislike of the play, but Bryan College professor (and Scopes Trial historian) Richard Cornelius suggested that he instead use the court transcripts to produce a historically accurate play about the Scopes Trial. In July, Chapin did just that; *Destiny in Dayton*, using local actors, premiered in the same courtroom in which Scopes was convicted 63 years earlier. In 1990, the play was supplemented with a festival featuring music, vendors, quilt shows, tours, traditional arts and crafts, and celebrities such as Miss Dayton, Strawberry Queen, and Mrs. Grandma Tennessee (Halbrook, 2012).

In 1998, the festival's play was changed to the longer *Monkey in the Middle*. That play, written by Dayton resident Gale Johnson, was more sympathetic to Dayton and Bryan, and tried "to give people not only a feel for the trial, but of the events taking place in 1925 and the attitudes of the people and the media" (Shackleford, 2002). In 2007, Johnson's play was recorded by Bryan College and sold as *Inherit the Truth*.

In 2000, to celebrate the 75th anniversary of the trial, the festival featured a symposium and a "reunion" of descendants of trial participants, including Judge Raulston's granddaughter, John Butler's grandson, and T.J. Brewer, the only still-living participant in the trial, who as a four-year-old sat on Judge Raulston's podium and pulled the names of potential jurors from a hat (Clayton, 2000). In 2008, the festival's play was replaced by Ted Kachel's performance of Bryan discussing "Bryan Reflects on the Scopes Trial's Significance for America."

From 2009 to 2011, the festival featured Dayton resident Curtis Lipps' play *One Hot Summer.* Lipps, like Johnson, tried to "tell the truth about the Scopes Trial" and present "the history that hadn't been taught." In 2012, the festival began featuring Deborah Harbin's *Front Page News* presented by Cumberland Playhouse. According to Scopes Trial expert Tom Davis, the festival's play continues "to give [attendees] the facts and let them make their decision. If you don't like Bryan, that's your privilege. But dislike him because of what the reality is, not because of *Inherit the Wind*" (Head, 2000).

Sources

Clayton, Ron. 2000. Symposium, reunion part of Scopes 75th anniversary. *Chattanooga Times Free Press,* July 13, B1.
Halbrook, Patrick Nason. 2012. The Scopes Trial in American Memory. Master of Arts Degree, Department of History, North Carolina State University.
Head, John. 2000. Southern almanac: Fest brings event to life. *Atlanta Journal and Constitution,* July 23, 2M.
Shackleford, Chris. 2002. Rhea festival will center on 1925 Scopes evolution trial. *Chattanooga Times Free Press,* June 20, B2.

Scopes Trial Trail

The Scopes Trial Trail is a set of plaques posted throughout Dayton that mark sites associated with the Scopes Trial. The plaques, which were posted in the mid–1990s,

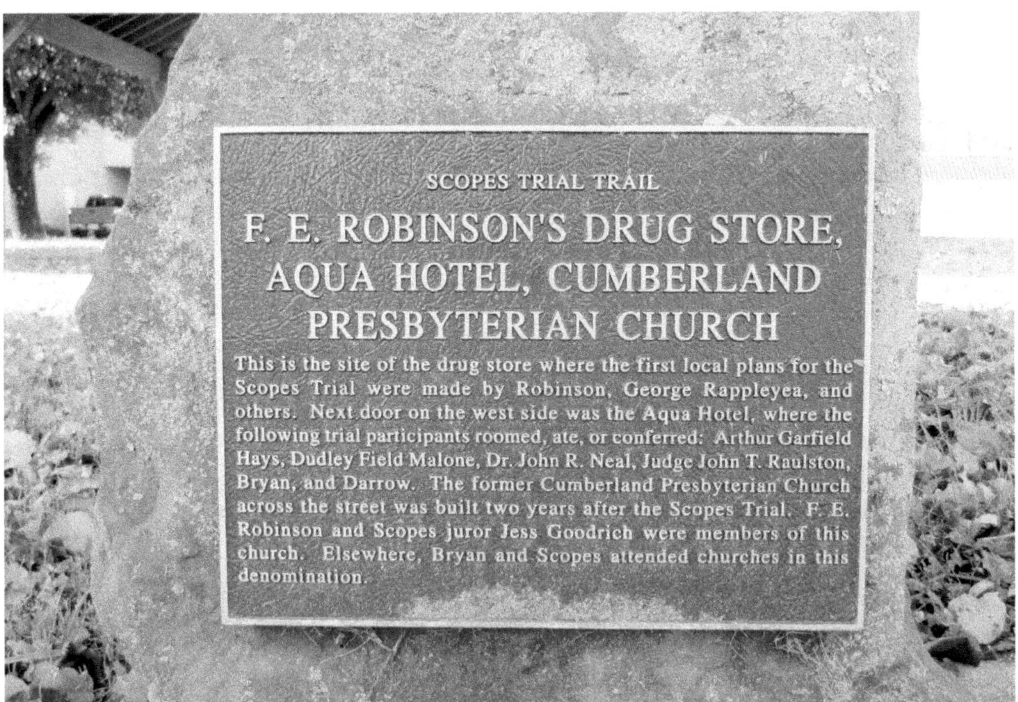

Today's visitors to Dayton can follow the Scopes Trial Trail, which commemorates important sites associated with the trial. This plaque marks the former site of Robinson's Drug Store (where the trial originated), and Hotel Aqua (where Judge Raulston and several other trial participants stayed) (Randy Moore).

designate Hotel Aqua and Robinson's Drug Store, Bailey Boardinghouse, Bailey Hardware, the Broyles-Darwin Home, Bryan College, Cedar Hill, First United Methodist Church, McKenzie Law Firm, Morgan Furniture Company, Rhea County Courthouse, Rhea Central High School, Smith's Crossroads, and the homes of A.P. Haggard, F.R. Rogers, Walter F. Thomison, and F.E. Robinson (Moore, 2018). The plaques were the brainchild of Richard Cornelius (1934–2014), a Bryan College professor and Scopes Trial expert.

Source

Moore, Randy. 2018. *A Field Guide to the Scopes Trial*. Dayton, TN: Rhea County Historical and Genealogical Society.

Shelton, Harry J.
1908–1994

Harry J. "Bud" Shelton was born on February 2, 1908, in Rome, Georgia. Seven years later, Shelton and his family moved to Dayton. In 1925, Shelton was a student in Rhea Central High School, where John Scopes taught him science. At the Scopes Trial, the 17-year-old Shelton testified that Scopes had told his class, "All forms of life begin with the cell." When Darrow then asked, "Did you leave the church when he told you all forms of life begin with a single cell?," Shelton replied, "No sir." After testifying, Shelton left the courtroom with fellow student Howard Morgan and asked, "Don't you think Bryan is a little narrow-minded?" (Linder, 2020).

Scopes encouraged Shelton and other students to testify against him. Shelton liked Scopes but admitted, "We were not concerned with evolution or Scopes. Our minds were on the thoughts of spring—baseball and fishing," later adding that Scopes' teaching "had no profound effect on me at all. In fact, I wasn't even interested in the subject" (Brown, 1985; Scopes Trial: Circus came to town, 1985). Shelton's mother wasn't concerned about evolution either, telling reporters, "As far as I'm concerned, they can teach my boy evolution every day of the year. I can see no harm in it whatever." Further easing Mrs. Shelton's mind was the fact that Harry had "forgotten most of his lessons" and "had to get the book out and study it up" before the trial (Linder, 2020). Shelton later claimed, "Scopes never taught me any evolution," adding that the information from the textbook that Scopes had covered "had already been taught by our regular teacher, Ferguson, before he got sick." To Shelton, the trial "was all a set-up" (Brown, 1997).

Shelton graduated from Rhea Central High School in 1927, after which he served as a chief warrant officer in the U.S. Army. Shelton then returned to Dayton, where worked for the R.E. Winsett Music Company and was active in a variety of civic organizations (Carpenter, 1994).

In 1990, at age 82, Shelton wrote to Arkansas schoolboy Jason Blankenship, "I don't think [the Scopes Trial] put our community in a favorable light and it opened a 'can of worms,' so to speak, as the controversy is still very much discussed and debated today" (Maxwell, 2014). Shelton, like other residents of Dayton, understood that Kramer's *Inherit the Wind* "had to entertain, but it was a gross exaggeration." In July 1986, Shelton had claimed, "I certainly don't believe that any of my ancestors hung by their tails—and do believe that man was created in the 'Image of God'" (Guffey, 2015).

Shelton, the last living student-witness in the Scopes Trial, died on June 15, 1994. He is buried atop a hill in Dayton's Rhea Memory Gardens.

Sources

Brown, Fred. 1985. Student recalls Scopes trial and famous teacher. *Daily Post Athenian* (Athens, TN), March 22, 3.
Brown, Rodger Lyle. 1997. *Ghost Dancing on the Cracker Circuit.* Jackson: University of Mississippi, 128.
Carpenter, John B. 1994. Bud Shelton, last Scopes witness, dies. *Dayton Herald-News*, June 16, 1A.
Guffey, Pat Hawkins. 2015. The Scopes Trial as remembered by H.J. Shelton. *Rhea Notes*, 22 (4), 3–5.
Linder, Douglas O. 2020. The student witnesses in the Scopes Trial. Accessed October 1, 2020, at https://-famous-trials.com/scopesmonkey/2394-the-student-witnesses-in-the-scopes-trial.
Maxwell, Angie. 2014. *The Indicted South: Public Critics, Southern Inferiority, and the Politics of Whiteness.* Chapel Hill: University of North Carolina Press, 30.
Scopes trial: Circus came to town. 1985. *Daily News* (Rutherford, TN), July 22, 1.

Stewart, Arthur Thomas
1892–1972

Arthur Thomas "Tom" Stewart was born on January 11, 1892, in Dunlap, Tennessee. After graduating in 1909 from the Pryor Institute, a private Methodist school in Jasper, Tennessee, Stewart enrolled in the University of Tennessee, but had to drop out after two months because of illness. He then attended Emory College (now Emory University) in Atlanta, and in 1913 earned a law degree from Cumberland College in Cumberland, Tennessee. After practicing law for two years in Birmingham, Alabama, Stewart returned to Jaspar and practiced law from 1915 to 1919. While in Jaspar, Stewart served as an assistant to Attorney General Ben McKenzie, with whom he would work again during the Scopes Trial.

From 1923–1939, Stewart served as District Attorney for Tennessee's Eighteenth Circuit. Stewart's most famous case was the Scopes Trial. As that trial approached, Stewart insisted that Scopes' prosecutors would "vindicate the two rights involved: (1) the right of the people to control the schools, and (2) the right of the parents to guard to religious welfare while in school." Stewart reminded reporters that "the public schools of the state are maintained by the funds collected from the taxpayers.... The state has a right, through its legislature, to control the curriculum set for the public schools" (Defines issues in Scopes case, 1925; Legal phase lost sight of, 1925).

During the trial, Bryan made the prosecution's case outside the courtroom, and Stewart made the case inside the courtroom. When Stewart learned that the defense would try to use scientists "to show that there is no conflict between the Bible and science because they claim that the evolutionary process was simply the method by which God created man" and that Scopes "will not deny that he taught in the high school of Rhea County that man descended from a lower order of animals," Stewart responded: "If this be true, [Scopes] is guilty of the violation of the law and all the evidence of scientists within his reach cannot save him. The law, itself, says that it shall not be taught to the students in the public schools that man descended from a lower order of animals. That is the issue and the only issue in this case: whether Scopes taught [evolution] or did not teach [evolution]. Scientists might claim that man descended from a lower order of animals by the process of God's creation, but this can make no difference. The law says such cannot be taught. If such evidence should be gone into by both sides, that is, if the defense should undertake to show that there is no conflict between the Bible and evolution by expert witnesses, and the state likewise to show by expert witnesses that there is a conflict, it would become a never-ending proposition and would in the end be but a babel of tongues" (Narrows scope of Scopes Trial, 1925).

Throughout the trial, Stewart argued that (1) the only issue before the court was whether Scopes had violated the Butler Act (i.e., not whether evolution or the Butler Act were valid) and (2) the curriculum of public schools should be set by the state legislature, not the courts. Stewart's strategy prevented Scopes' defenders from calling scientists as witnesses and questioning the validity of evolution. To Stewart, Scopes' case was simple: The Tennessee legislature had passed the Butler Act and John Scopes had violated it. Stewart had several heated confrontations with Scopes' defenders, including one on July 16 in which Stewart proclaimed, "In the name of the great God, I am on the side of religion.... I stand with religion because I want to know beyond this world that there may by an eternal happiness for me and for all.... [Darrow] is the greatest criminal lawyer in America today. His courtesy is noticeable—his ability is known ... great God, the good that a man of his ability could have done if he had aligned himself with the forces of right instead of aligning himself with those which sink their fangs into the bosom of Christianity" (Clapper, 1925).

When Hays called Bryan as a witness on July 20, Stewart objected. After his objection was overruled, Stewart again objected, claiming that Darrow's questioning had "gone beyond the pale of any issue that could possibly be injected into this lawsuit, except by imagination. I do not think the defendant has a right to conduct the examination any further and I ask Your Honor to exclude it." But Raulston disagreed, simply saying, "I will hear Mr. Bryan.... Mr. Bryan is willing to be examined. Go ahead." Later in the questioning, Stewart again objected, noting that "we have left all annals of procedure behind. This is a harangue.... It is not worth anything before a jury." Stewart then pleaded, "This has gone beyond the pale of a lawsuit, Your Honor. I have a public duty to perform, under my oath and I ask the court to stop it." Stewart's objections were again ignored.

Except for Darrow's questioning of Bryan, Stewart controlled the Scopes Trial. Stewart was praised for his work, for a time even by Mencken. Years later, Stewart admitted that he "felt rather awed in the presence of such men as Mr. Bryan, Clarence Darrow, who joined the defense counsel, and other lawyers of national fame" involved in the trial (Stewart rites set tomorrow, 1972).

Stewart's son later claimed that his father knew that Judge Raulston's mistake during sentencing would reverse Scopes' conviction (Olasky and Perry, 2005). In 2018, Stewart's grandson, Jeffrey Stewart, told reporters that his grandfather's original argument to Judge Raulston noted that there was "no punishment for expert testimony. If you don't like the first expert, you just keep calling others until they agree with your opinion" (Tucker, 2018).

In 1938, Stewart won a special election to fill the vacancy created by the death of U.S. Senator Nathan Bachman (1878–1937). Stewart could have been seated in the Senate immediately, but he waited until January 1939 so he could finish his term as District Attorney. In 1942, Stewart was easily reelected (with 69 percent of the vote) when he defeated Republican F. Todd Meachem (22 percent of the vote) and Scopes' defender and Independent Party candidate John Randolph Neal (9.6 percent of the vote). Soon thereafter, at the beginning of Japanese interment, the politically conservative Stewart introduced a bill to revoke the citizenship of Japanese-Americans.

In the 1948 Democrat primary election, Stewart was defeated by progressive Representative Carey Estes Kefauver (1903–1963). (John Randolph Neal, another Democrat candidate, won 0.5 percent of the vote.) Kefauver went on to win the general election (with 65 percent of the vote) and Stewart returned to Tennessee to practice law. Years later, Stewart was asked by filmmaker Stanley Kramer to be an advisor during pre-production of *Inherit the Wind*. When Stewart refused, Kramer made his character insignificant in the movie (Olasky and Perry, 2005).

The 80-year-old Stewart died of a heart attack in his home in Nashville, Tennessee, on October 10, 1972. He is buried in Memorial Park Cemetery in Winchester, Tennessee.

Sources

Biographical Directory of the United States Congress. 2020. Arthur Thomas Stewart. Accessed September 21, 2020, at https://bioguideretro.congress.gov/Home/MemberDetails?memIndex=s000901.
Clapper, Raymond. 1925. Argues to bar scientists in Dayton trial. *Des Moines Register*, July 17, 1.
Defines issues in Scopes case. 1925. *Chattanooga News*, June 23, 1.
Legal phase lost sight of. 1925. *Nashville Banner*, June 17, 1.
Narrows scope of Scopes Trial. 1925. *Chattanooga Daily Times*, July 11, 1.
Olasky, Marvin, and John Perry. 2005. *Monkey Business*. Nashville: Broadman & Holman, 138, 331, n174.
Stewart rites set tomorrow. 1972. *Nashville Tennessean*, October 12, 1, 17.
Tucker, Becky. 2018. Stewart shares family's memories of historic trial. *Dayton Herald-News*, July 15, 1–2.

Straton, John Roach
1875–1929

John Roach Straton was born on April 6, 1875, in Evansville, Indiana, into a Baptist pastor's family. When he was 18, Straton—who then embraced humanism and evolution—went to Atlanta, Georgia, to study law. There, Straton converted to Christianity, graduated from Mercer University in 1898, and was ordained as a minister two years later. Straton was an exceptional debater, earning two gold medals for winning the Georgia and Southern Intercollegiate Oratorical Championships. After enrolling in Southern Baptist Theological Seminary (Louisville, Kentucky) from 1902 to 1903, Straton was hired as a literature professor at Baylor University.[47] He then preached in Chicago (1905–1908), Baltimore (1908–1913), and Norfolk, Virginia (1914–1917) before moving in May 1918 to New York City's Calvary Baptist Church, where he served until his death. When Straton's fiery sermons and public confrontations of sinners in bars and brothels began making headlines, writers labeled him "The Fundamentalist's Pope" and "The Witch Doctor of Gotham." Straton was ambivalent about the KKK; he sometimes disowned the organization, but at other times voiced approval of their principles (Straton, 1968).

The militant Straton—a charter member of the WCFA—first denounced the "moral failure of evolution" in August 1918, and by the early 1920s he led the fundamentalists' antievolution campaign in the northeast. He publicly feuded with Henry Fairfield Osborn, whose American Museum of Natural History included displays that were "treason to God almighty" because they were "poisoning the minds of school children by false and bestial theories of evolution" (Pulpits continue doctrinal dispute, 1924). Everywhere Straton spoke, headlines such as "Straton Launches War on Evolution" and "Monkey Theories the Root of Evil" followed (Straton launches, 1928).

Straton, who was one of the first ministers to preach a sermon via radio directly from his pulpit, told overflow crowds at his church, "The great battle of the age is now on between Christianity and evolution … it is better to wipe out all the schools than undermine the Bible by permitting the teaching of evolution" (Straton, 1925b). According to Straton, evolution's "degrading philosophy of animalism" had produced women's vulgar clothes, "the appalling divorce record," "cabaret orgies" of theater, dance, literature "reeking with moral infamy," "the widest wave of immorality in the history of the human race," "the degradation of womanhood," and even "the substitution of dogs for babies" (Straton wins evolution debate, 1924; Straton, 1920, Says evolution hits standards, 1924). Straton sinisterly asked, "We would visit summary execution upon a teacher who poisoned a student and thus killed his body. What ought we to do then to a teacher who poisons the mind with untruth, and thus kills his soul?" (Straton, 1918).

In 1923–1924, Straton confronted Kirtley Mather, Maynard Shipley, and fundamentalist-turned-Unitarian preacher Charles Francis Potter in a series of debates, each heard by more than 3,000 spectators at Carnegie Hall (across the street from his church) and tens of thousands more on Calvary Baptist's radio station (WQAO, the first church-owned-and-operated radio station in the United States). In the debates, which were reported in newspapers across America, Straton stressed that someone could not be both a Christian and an evolutionist because Christianity and evolution are "mutually exclusive. If it is Christianity, then it is not evolution; and if it is evolution, then it is not Christianity" (Straton, 1925a). To Straton, the stakes were high because "the wave of immorality" threatening children "is easily traced … [to] when the dark and sinister shadow of Darwinism first fell across the fair fields of human life.… If we are merely highly developed beasts, then why should we not live like beasts? Monkey men make monkey morals.… America's educational system ultimately will be wrecked if the teaching of evolution is allowed to continue" (Straton, 1925a). Straton likened Biblical creationism to "monogamy, the sacredness of the marriage vow … and the sanctity of the home." Just before the Scopes Trial began, Straton likened Bryan to Christ, and Darrow to Satan.

Bryan implored Straton to come to Dayton to help rebut the testimonies of the defense team's witnesses. Straton promised to be there "at the drop of a hat" and "put up a man-sized fight" against evolution's "sin and godlessness," but he stayed home and wrote columns for newspapers instead (Giordano, 2008; Straton, 1925a). When Bryan died five days after the trial, the publicity-savvy Straton—repeatedly claiming, "Everywhere I have been, I have been urged to take up Mr. Bryan's work"—was the first of many fundamentalists who declared themselves to be Bryan's God-appointed successor (Dr. J. Roach Straton condemns, 1925).

After the Scopes Trial, Straton repeatedly challenged Malone and Darrow to debates, but was rebuffed. In an October 1925 speech at Harvard University, Straton condemned Kirtley Mather (a Baptist deacon and professor there), dismissed evolution as a "fetish … [that] takes the place of God," and denounced Darrow, Mencken, Malone, and all other "atheists and advocates of the degenerate cult" of evolution. According to Straton, evolution had produced "the greatest wave of degeneracy and crime that has swept the world since the fall of the Roman Empire" (John Roach Straton, 1925a).

In 1927, Straton began promoting for The Supreme Kingdom, an antievolution organization organized by Edward Young Clarke. (Clarke, a deposed Imperial Wizard who had been booted from the KKK for violating the Mann Act, had often converted religious activities into personal wealth.) Clarke promised Straton $30,000 ($500 for each of 60 lectures, which dwarfed Stratton's annual salary of $12,000 at Calvary Baptist Church) to speak and recruit members for the Kingdom. However, a series of financial scandals soon forced Straton to resign (e.g., in addition to paying himself an annual salary of $25,000, Clarke was keeping 50–75 percent of the membership dues for himself). To recover money lost in the scandals, Clarke announced that The Supreme Kingdom (unlike the Klan) would open its membership to Catholics, provided they were Christian, White, and had $10 (Macon newspaper, 1927; New Straton cult, 1927). Despite this and other changes, Clarke's organization soon disappeared.

In April 1929, Straton suffered a stroke that left him partially paralyzed. A nervous breakdown followed, and on October 29, 1929—that is, "Black Tuesday," the day on which the stock market crashed—Straton died of a stroke in a sanitarium at Clifton Springs, New York. He is buried in a family plot in Warwick, New York, beneath the epitaph "A Good Soldier for Jesus Christ."

Sources

Dr. J. Roach Straton challenges Darrow. 1925. *Chicago Daily News*, August 18, 1.
Giordano, Ralph G. 2008. *Satan in the Dance Hall: Rev. John Roach Straton, Social Dancing, and Morality in 1920s New York City*. Lanham, MD: Scarecrow Press, 150.
John Roach Straton condemns evolution as degenerate cult. 1925. *The Harvard Crimson*, October 16.
Macon newspaper "calls" out Straton. 1927. *The (Columbia, SC) Sunday Record*, January 9, 1, 10.
New Straton cult patterned on Klan. 1927. *New York Daily News*, January 19, p. 12.
Pulpits continue doctrinal dispute. 1924. *New York Times*, March 9, 20.
Says evolution hits standards. 1924. *Lexington (Kentucky) Leader*, September 26, 10.
Straton, Hillyer H. 1968. John Roach Straton and the KKK. *Andover-Newton Quarterly*, 9, 124–134.
Straton, John Roach. 1918. Choosing an education, or the moral failure of evolution. Typed synopsis of a sermon to students preached on August 4, 3.
______. 1920. *The Menace of Immorality in Church and State*. New York: George H. Doran.
______. 1925a. *The Famous New York Fundamentalist-Modernist Debates*. New York: George H. Doran.
______. 1925b. *The Old Gospel at the Heart of the Metropolis*. New York: George H. Doran.
Straton launches war on evolution. 1928. *New York Times*, April 16, 25.
Straton wins evolution debate: 3,000 cheer. 1924. *Searchlight*, 7, 12 (February 8), 4.

Sunday, William Ashley 1862–1935

William Ashley "Billy" Sunday was born on November 19, 1862, just outside of Ames, Iowa. His father died of pneumonia during the Civil War, and Billy's impoverished mother later sent 10-year-old Billy and an older brother to the Iowa Soldiers' Orphans' Home. Sunday never graduated from high school, but was a good athlete. In 1883, Sunday's speed and agility produced what would become an eight-year career in professional baseball, during which he was an average hitter (.248 over 499 games), a good fielder, and an exceptional base-stealer (Knickerbocker, 2000).

In 1886, Sunday converted to Christianity and, four years later, turned down $3,500 per year to play baseball for a salary of $83 per month to work at the Chicago YMCA. By the early 1900s, Sunday's colloquial, frenetic "circus salvation" services were attracting large crowds as he equated the teaching of evolution with prostitution, crime, alcohol, and anarchy. Sunday screamed from atop his pulpit, smashed chairs while battling Satan, and ran across the stage to slide safely into "home" in heaven. Sunday then promised his followers that their work to rid society of evolution and alcohol would result in "the great Umpire of the Universe" proclaiming them "safe at home" (Billy Sunday, 2020).

Sunday defended his unorthodox, colloquial style by telling crowds, "I want to preach the gospel so plainly that men can come from the factories and not have to bring a dictionary," often adding, "I know no more about theology than a jackrabbit does about ping-pong, but I'm on my way to glory" (Billy Sunday, 2020). Sunday's blunt sermons about "the sex question" were described as "the ugliest, nastiest, most disgusting" ever heard from a preacher (Betts, 1916). For weeks at a time during some of his revivals, Sunday proclaimed, "I don't believe the old bastard theory of evolution, either. I believe it's pure jackass nonsense. When the consensus of scholarship says one thing and the Word of God another, the consensus of scholarship can go plumb to hell for all I care.... If you believe your great, great grand-daddy was a monkey, then you can take your daddy and go to hell with him!" (Billy Sunday hits, 1925; Simons and Kerr, 1915). Sunday often confronted religious hypocrites with claims such as "Going to church doesn't make you a Christian any more than going to a garage makes you an automobile." As one reporter noted, Sunday's sermons "left [worshippers] thrilled and bewildered as they have never been before" (Billy Sunday, 2020).

Sunday preached that evolution was endorsed by "godless bastards and godless losers," that any preacher who accepted and taught evolution is "a stinking skunk, a hypocrite, and a liar" who could "go to hell," and that Charles Darwin was burning in hell (Bruns, 1992). Sunday wanted all evolution teachers to be fired (e.g., "stand [them] up before a firing squad") because Christianity could never be reconciled with the "God-forsaken, hell-born, bastard theory of evolution," and anyone who claimed otherwise was a "theological bootlegger" who could "go to hell." Sunday, who proudly announced that his ancestors "didn't hang from trees by 'prehensile' tails," grouped Darwin, Darrow, and evolutionists with "atheists and the infidels and the whoremongers," all of whom could "go to hell" (Higgs, 1995). Sunday told many people to go to hell.

In early 1925, while the Tennessee legislature debated the Butler Act, Sunday led an 18-day revival in Memphis. Throughout the revival, which attracted more than 200,000 people (almost 10 percent of the people in Tennessee), Sunday urged Tennessee's legislators to pass the bill and save their state from "the God-forsaken gang of evolutionary cutthroats." The legislators and governor responded by passing the Butler Act, which criminalized the teaching of human evolution in Tennessee's public schools. Just before the Scopes Trial began, Sunday reassured his followers that "God will eventually triumph over the gorilla" (Billy Sunday wallops the devil, 1925).

Sunday's preaching throughout the Midwest, and especially in Tennessee in early 1925, helped energize the fundamentalist movement, secure the passage of the Butler Act, and produce the Scopes Trial. Scopes' prosecutor William Jennings Bryan and Rhea County School Superintendent Walter White begged Sunday to come to Dayton for the trial, but he did not show up, instead sending Bryan a note reassuring him that "all the believing world is back of you in your defense of Christ and the Bible" (Sunday to Bryan, 1925).

Sunday was the most famous, influential, celebrated American evangelist of his time. Throughout his career, the rags-to-riches Sunday—a revered and reviled icon of the Roaring Twenties—led more than 300 revivals and spoke (without radio or microphones) to more than 100 million people. No one else, either before or since, has spoken directly to so many people. Sunday popularized revivalism; famous subsequent preachers such as Jim Bakker, Oral Roberts, Billy Graham, and Jimmy Swaggart emerged from Sunday's revivalist tradition.

Sunday's popularity later waned as people discovered radio, movies, and other forms of entertainment. After having a mild heart-attack in early 1935, Sunday suffered a fatal attack at his brother-in-law's home on November 6, 1935, in Chicago, Illinois. More than 3,500 people attended Sunday's funeral at Chicago's Moody Church, at which the most prominent floral arrangement came from the Chicago Cubs. Memorial services across the country mourned Sunday's death. Sunday is buried in Forest Home Cemetery in Forest Park, Illinois, beneath the inscription from II Timothy 4:7: "I have fought a good fight, I have finished my course, I have kept the faith."

Sources

Betts, Frederick William. 1916. *Billy Sunday: The Man and Method.* Boston: Murray Press.

Billy Sunday. Salty evangelist. Accessed April 19, 2020, at https://www.christianitytoday.com/history/people/evangelistsandapologists/billy-sunday.html.

Billy Sunday hits modernism and wets. 1925. *New York Times*, December 21, 15.

Billy Sunday wallops the devil and gorillas. 1925. *Denver Post*, July 1, 19.

Bruns, Roger A. 1992. *Preacher: Billy Sunday and Big-Time American Evangelism.* Urbana: University of Illinois Press.

Higgs, Robert J. 1995. *God in the Stadium: Sports and Religion in America.* Lexington: University of Kentucky Press, 225.

Knickerbocker, Wendy. 2000. *Sunday at the Ballpark: Billy Sunday's Professional Baseball Career, 1883–1890.* Lanham, MD: Scarecrow Press.

Simons, Algie, and Charles H. Kerr. 1915. Looking 'em over. *International Socialist Review*, 16 (8), 465–466/

Sunday, Billy, to William Jennings Bryan, July 4, 1925. William Jennings Bryan Papers, Library of Congress, Box 47.

Thomison, Walter Fairfield 1859–1940

Walter Fairfield Thomison was born on October 14, 1859, in Washington, Tennessee. After graduating from Vanderbilt University Medical School around 1895, Thomison became the physician for Dayton Coal and Iron Company. In 1907, Thomison opened Dayton's first drugstore, next to Hotel Aqua.

On the afternoon of July 26, 1925, Thomison was called by neighbor A.B. Andrews to the home of Andrews's neighbor, F.R. Rogers, which William Jennings Bryan had rented during the trial. Bryan, who had laid down for a nap, was unresponsive, and Thomison could not revive him. Thomison pronounced Bryan dead at 3:30 p.m. of apoplexy (hemorrhage or stroke).

In 1929, Thomison and his son Agnew Thomison opened Cedar Hill, Dayton's first hospital. In the 1930s, they moved their hospital to the rooms above Rogers's Pharmacy on Market Street. That hospital, whose sign remains near 1325 Market Street, closed in 1960. Thomison, who was the surgeon for Southern Railway for 47 years, died on May 5, 1940, and is buried in Dayton's Buttram Cemetery.

Source

Broyles, Bettye J. 1991. *History of Rhea County, Tennessee.* Dayton, TN: Rhea County Historical and Genealogical Society.

Ussher, James (1581–1656)

James Ussher was born into a merchant family in Dublin, Ireland, on January 4, 1581. At age 13, he was one of the first students to enroll in Dublin's Trinity College and, in 1625, was named archbishop of Armagh (a town in Northern Ireland). In 1650, after more than 10 years of studying Biblical genealogy, Ussher announced in the first paragraph of his *Annals of the Old Testament* that God began creating heaven and earth "upon the entrance to the night preceding the twenty-third day of Octob[er], in … the year before Christ 4004.…" Ussher's date, which did not include a time of day, became the most famous of the many "creation dates" (e.g., calculated by Martin Luther, Johannes Kepler, Isaac Newton, and others) when, in 1701, the Church of England began including it adjacent to Genesis 1:1 in its official Bibles. For nearly 300 years, most King James Versions of the Bible included Ussher's date. It also appeared elsewhere, such as in Shakespeare's *As You Like It*, in which Rosalind laments, "The poor world is almost 6,000 years old." The Bible introduced at the Scopes Trial included Ussher's famous date (Moore and Decker, 2008). Ussher's chronology is often ridiculed today, but at the time it was a remarkable scholarly publication.

During the Scopes Trial, when Darrow asked when the flood occurred, Darrow prompted Bryan with "About 4004 BC?," to which Bryan responded, "That has been the estimate of a man [i.e., Ussher] that is accepted today. I would not say it is accurate." When Darrow later returned to the topic, Bryan said, "Let me see Usher's [sic] calculation about it." After more questions, Darrow asked, "Have you any idea how old the earth is?" and again prompted Bryan with "It says B.C. 4004." Bryan answered, "That is Bishop Usher's [sic] calculation."

Stewart claimed that Darrow's questioning of Bryan was inappropriate and pleaded for Judge Raulston "in the name of all that is legal, to stop this examination and stop it here." Hays then pointed out that "Mr. Bryan has already stated that the world is not merely 6,000 years old," to which Bryan responded "You think the Bible says that." When Hays claimed "The one [i.e., Bible] you have taken in evidence says that," Bryan announced, "I don't concede that it does."

Ussher's "4004" date appears five times in the court transcript; first in the written testimony of Kirtley Mather, next in the written testimony of Winterton Curtis,[48] twice by Darrow when questioning Bryan, and finally by Hays during the questioning of Bryan (all in the July 20 transcript). Ussher's date was also mentioned indirectly by Ben McKenzie on July 16.

Ussher died on March 21, 1656, and was buried in St. Paul's Chapel in Westminster Abbey.

Source

Moore, Randy, and Mark Decker. 2008. *More than Darwin: An Encyclopedia of the People and Places of the Evolution-Creationism Controversy*. Westport, CT: Greenwood Press.

Weather

On Monday, July 20, Judge John Raulston moved court proceedings to an outdoor dais because he worried that the floor of the courtroom was "heavily burdened" and unstable. Mayor A.P. Haggard disagreed with Raulston's assessment, telling reports that the floor "is safe ... there is nothing at all the matter with it." Scopes had a simpler explanation; although Scopes admitted that Raulston's claim "made a good story," Scopes knew "the fact was otherwise: It was the extreme heat that shifted the Court outdoors." As Scopes noted, Dayton was "as hot as any hell the white-headed evangelist, T.T. Martin, might have conjured up" (Scopes and Presley, 1967).

Here are the weather reports from the U.S. Weather Bureau published in the *Chattanooga Daily Times* on July 11, 14, 15, 16, 17, 18, 21, and 22. The heat index, which combines temperature and relative humidity, is what the temperature feels like to humans.

Date	*Day*	*Temperature (°F)*		*Relative Humidity, %*	*Heat Index*
		High	*Low*		
July 10	Friday	86.0	70.2	59	91
July 13	Monday	96.0	73.2	44	103
July 14	Tuesday	94.7	73.2	36	96
July 15	Wednesday	93.2	74.3	52	102
July 16	Thursday	97.2	72.3	33	99
July 17	Friday	85.1	71.3	33	83
July 20	Monday	92.0	69.9	35	92
July 21	Tuesday	75.2	69.2	36	74

The hottest day of the trial was Thursday, July 16, when the temperature reached 97.2°F, and the day that felt the hottest was Monday, July 13, when the heat index reached 103°F. Monday, July 20, when the trial moved outside, was and felt like the fifth-hottest day (of eight) of the trial.

Source

Scopes, John T., and James Presley. 1967. *Center of the Storm: Memoirs of John T. Scopes*. New York: Holt, Rinehart and Winston, 101, 167.

WGN
Est. 1924

WGN is a commercial AM radio station in Chicago, Illinois. It was purchased in 1924 (as WDAP) by the *Chicago Tribune*, whose "World Greatest Newspaper" slogan became the basis for WGN's call-letters. Soon after WGN began broadcasting, WGN's Quin Ryan started anchoring live broadcasts of music, comedy, sporting events, and political debates, including one about Prohibition involving Clarence Darrow. In Chicago, WGN transmitted from the Drake Hotel (Calibraro and Fink, 1961).

On June 28, 1925, the *Chicago Tribune*—whose editorial board had denounced Bryan's opposition to the teaching of evolution (WGN, 1925)—announced, "WGN Will Bring Evolution Case to Your Home." It did: WGN's broadcast of the Scopes Trial in July 1925 was the first live broadcast of a criminal trial in American history.[49] WGN was given permission to broadcast the trial by Scopes' prosecutor and Rhea County Judge,

Gordon McKenzie, who was custodian of the courtroom and issued all of the trial's press-passes. WGN rearranged the courtroom to suit its placement of four microphones: one in front of the witness stand, one in front of the defense team's table, one in front of the prosecution team's table, and one that stayed with Ryan, both in the courtroom and in a small, quiet, nearby cloakroom from which he provided commentary. In return for broadcasting privileges of the now made-for-radio trial, WGN agreed to place four loud-speakers at each of five locations: the courthouse lawn, a grammar school, RCHS, the American Legion hall, and the tabernacle. The newness of radio, as well as the public's interest in the trial, produced crowds at each site and made Ryan a celebrity in Dayton. People were fascinated by "the little box that picks sounds from the air" (Wesolowski, 1975; Calibraro and Fink, 1961). When Rhea County School Superintendent Walter White proclaimed his belief in Biblical miracles, he added, "None of [those] stories are as hard to believe as radio" (Sumner, 1925).

WGN began installing its equipment in Dayton on July 8, two days before proceedings began. Technical problems disabled the broadcast on the trial's first day, so live broadcasts did not start until Monday, July 13. This meant that the audience never heard jury selection or Scopes' indictment. To broadcast the trial, WGN paid $1,000 per day to rent telephone cables stretching the 540 miles from Dayton to Chicago.

During the trial, the jury was seldom in the courtroom. When they retired to the courthouse lawn, they could have listened on WGN's speakers to the proceedings that had just been ruled inappropriate for them to hear. Realizing that this might happen, Judge Raulston instructed the jury not to "linger in the courthouse yard." Jury foreman Jack Thompson assured Raulston, "There is not a single juryman that has heard a single word pass over the horns out there." The station's total cost for broadcasting the trial (including Ryan's salary of $60 per week) was less than $10,000 (Wesolowski, 1975).

WGN, now owned by Nexstar Media, continues to broadcast in Chicago. In 2000, WGN celebrated the 75th anniversary of the Scopes Trial by broadcasting a reenactment of the trial during Dayton's Scopes Trial Play and Festival. In the reenactment, WGN's John Williams played the role of Quin Ryan, his real-life predecessor.

Sources

Calibraro, Daniel D., and John Fink. 1961. *WGN: A Pictorial History*. Chicago: WGN.
Sumner, Allene M. 1925. Adam and Eve lived. *Knoxville News*, June 23, 7.
Wesolowski, James Walter. 1975. Before Canon 35: WGN broadcasts the Monkey Trial. *Tennessee Historical Quarterly*, 34 (4), 392–406.
WGN. 1947. *A Century of Tribune Editorials*. Chicago: Chicago Tribune, 95–97.

White, Walter
1880–1951

Walter White was born on December 24, 1880, in Meigs County, Tennessee. After attending public schools there and in Rhea County, he attended the University of Chattanooga and other universities. White, who promised voters free school-books published by the state, $3 auto tags, "old age pensions," and "no saloons or whisky under any conditions," served several terms in the Tennessee House of Representatives, followed by seven terms as Rhea County School Superintendent. When the Butler Act passed in 1925, White—who had by then been superintendent for more than six years—thanked

Governor Peay for "the greatest step that Tennessee has taken since the saloon was abolished. You deserve the thanks of all honest people" (Walter White to Austin Peay, 1925).

On May 5, 1925, White went to the office of city attorneys Herbert and Sue Hicks and asked if Rhea County's schools should be concerned with the newly passed Butler Act. Sue Hicks "told him to ignore the law … [because] it might be declared unconstitutional" (Allem, 1959). Sue Hicks and White continued to talk as they walked to Robinson's Drug Store, where they met F.E. Robinson, George Rappleyea, and others who were daily customers there. Rappleyea showed the group the morning edition of the *Chattanooga Daily Times*, highlighting an ad that had been placed by the ACLU seeking a teacher to test the Butler Act. The businessmen, believing that a test-case could boost the area's struggling economy, allegedly asked biology teacher (and principal) W.F. Ferguson if he would be willing to test the law, but he declined.[50] They then summoned coach John Scopes, who was playing tennis nearby with some of his former students. When asked if he'd be willing to be arrested and test the law, Scopes agreed because, in his words, "somebody had to do it [i.e., test the law] … I was free, white, and 21 with no responsibilities, so I said, 'Go ahead, have at it'" (12 Days in Dayton, 2001). Rappleyea swore out a complaint, but when Justice of the Peace A.C. Benson admitted that he didn't know how to complete the warrant, Sue Hicks completed and signed the form himself (Allem, 1959; Glenn, 1975). When White left Robinson's store, he met the Rev. A.C. Stribling, a local correspondent for the *Chattanooga News*, whom he told, "Something has happened here that's going to put Dayton on the map!" (Allem, 1959). Soon thereafter, while raising money to prosecute Scopes, White began telling donors, "We represent the church and the public schools" (Many scientists in Scopes Trial, 1925; Prosecution asks financial aid, 1925). In the meantime, Benson told reporters that Scopes "is at least technically if not intentionally guilty" (Drugstore talk, 1925).

Rappleyea, who endorsed evolution, was Scopes' original prosecutor, but he was soon replaced by White, who was more sympathetic to the fundamentalists' cause. White dismissed Rappleyea's concern that the Butler Act was a religious law, asking "Isn't the Bible the basis of all truth?" White asked evangelist Billy Sunday to come to Dayton to help the prosecution, but Sunday—like all other prominent fundamentalist leaders—declined.

In mid–June, when Scopes announced to reporters that he was a Christian, White assured the public, "Under no circumstances will I ever recommend any person [for a teaching position] who is not a fundamentalist" (Scopes tells of religious beliefs, 1925). When the Scopes Trial opened on July 10, White described the upcoming events as "the battle of the century" ("Better give up schools than Bible," Bryan says, 1925).

On July 12, after hearing that some northern schools would reject students from Tennessee who had not been taught evolution, White announced that Dayton would build a fundamentalist college and invite Bryan to become its president (Want Bryan, 1925). In Dayton, White introduced Bryan to the school board as "the world's greatest man and its first citizen" (Bryan talks, 1925). White added, "Education is a woeful failure the world over until every school child believes in a real Adam and Eve, a real serpent who temped Eve, and a real angel with a flaming sword who drove the guilty twain from the Garden of Eden. Taxpayers should arise in protest that their hard-earned dollars are paid to teachers who tell their children that their grandparents are animals.… I believe that a real Adam and Eve were made by a man-like God and put in the Garden of Eden.… I must believe every Biblical word literally because if I question one word, I will question more—and then where will I be?" He then condemned Scopes, whom he had hired the previous summer, as "a menace to our youth, one who must pass through the fire of purging.... What have I or these little children left in life if men like Scopes teach us such things?" White, who "believe[d] in science as long as it is in harmony with the Bible," ended the interview by emphasizing humans' place in nature; apparently not a pet-lover, White reemphasized, "I cannot believe in a God who would make me in his image and that of an animal. I want my heaven, and I can't have it with dogs and cats" (Moutoux, 1925; Adam and Eve lived, 1925).

On July 15, White became the first witness to testify in the Scopes Trial. White, who had joked that he would "not insist on the death penalty for Scopes," testified that Scopes had taught science at Rhea Central High School and had admitted on May 5 in Robinson's Drug Store that he had "taught [from Hunter's *A Civic Biology*, the state-approved textbook] and had reviewed the entire subject ... [and] said he could not teach [that book] without teaching evolution." White did not claim that Scopes had taught his students about evolution, but he did claim that if Scopes had covered everything in the textbook, he must have taught evolution, because evolution is in the book. White did not testify again in the trial, but for the rest of his life, his business cards and stationery reminded people that he was "Prosecutor in Case of State of Tennessee v. John Thomas Scopes." He was proud that he had made Rhea Central High School "the most widely known high school in the United States" (Many science teachers seek Scopes' position, 1925).

On the trial's last day, the *New York Times* reported that White had asked U.S. Congressman J. Will Taylor (1880–1939; R-TN) to sponsor a federal antievolution bill banning any school that teaches evolution from receiving federal funds. Taylor admitted, "although I am a fundamentalist, I believe such a bill will do more harm to the church and Christianity than it would good. I will not sponsor any antievolution bill" (Taylor refuses to father bill, 1925).

In 1926, White ran for governor, but was soundly defeated by incumbent Democrat Austin Peay. White remained active in the Tennessee Republican party, and in 1940 was a delegate to the Republican National Convention in Philadelphia.

The 70-year-old White died of heart failure at his home in Dayton on February 15, 1951. He is buried in Buttram Cemetery in Dayton.

Sources

Adam and Eve lived. 1925. *Knoxville News*, June 23, 7.

Allem, Warren. 1959. *Backgrounds on the Scopes Trial at Dayton, Tennessee*. Master's Thesis, University of Tennessee, Knoxville, Tennessee, 61.

"Better give up schools than Bible," Bryan says on eve of battle. 1925. *Chattanooga Daily Times*, July 10, 1.

Bryan talks to school board. 1925. *Nashville Banner*, July 9, 2.

Drugstore talk started state's evolution case. 1925. *Nashville Tennessean*, May 24, 1.

Glenn, Juanita. 1975. Judge still recalls "Monkey Trial"—50 years later. *Knoxville Journal*, July 11, 17, 18, 30.

Many science teachers seek Scopes' position. 1925. *Chattanooga Daily Times*, July 19, 1.

Many scientists in Scopes Trial. 1925. *Chattanooga News*, June 26, 15.

Moutoux, John T. 1925. Prosecutor of Scopes has a big problem. *Knoxville News*, June 2, 9.

Prosecution asks financial aid in Scopes case. 1925. *Knoxville Journal*, June 7, 13.

Scopes tells of religious beliefs. 1925. *Nashville Banner*, June 17, 1.

Taylor refuses to father bill of Bryan group. 1925. *Chattanooga Daily Times*, July 21, 1.

12 Days in Dayton: The Scopes Monkey Trial. 2001. WTCI, Greater Chattanooga Public Television Corporation, 11:18.

Walter White, of Rhea, dead. 1951. *Chattanooga Times*, February 15, 1, 5.

Walter White to Austin Peay. April 15, 1925. Peay Papers, Box 43, Folder 1, Tennessee State Library and Archives, Nashville, TN.
Want Bryan to head a new Dayton college: Fundamentalists talk of national co-educational institution based on Bible. 1925. *New York Times*, July 13, 4.

Wright, George Frederick 1838–1921

George Frederick Wright was an American Congregational theologian and amateur geologist born into a farming family in Whitehall, New York, on January 22, 1838. He earned a A.B. from Oberlin College in 1859, an M.A. from the Oberlin Theological Seminary in 1862, and a D.D. from Brown University and a LL.D. from Drury College in 1887. He served in the Union Army in 1861 and later pastored Congregational churches in Vermont (1862–1872) and Massachusetts (1872–1881). After becoming a professor at Oberlin in 1881, he was made Professor of Harmony of Science and Revelation the following year.

Early in his career, Wright—a friend and protégé of Darwin's confidante Asa Gray (1810–1888)—was a "Christian Darwinist" (i.e., a theistic evolutionist) who reconciled evolution with orthodox Christianity. He was convinced that God had directly intervened in nature (e.g., "the Deity puts his hand directly to the work ... to supplement natural selection"), especially regarding the origin of humans and "the special endowments of man's moral nature." Wright, who restricted natural selection to "lower" plants and animals while making humans exempt from the process (Numbers, 1988), defended Darwin while attacking Darwinists who had made "illogical conclusions" when they replaced a designing deity with time and chance. Wright also wanted to "remove the misleading dates of Archbishop Usher [sic] from the margins of our 'authorized' translation [of the Bible] ... it is already high time this was done" (Wright, 1882).

By the late 1890s, Wright had become a fundamentalist and began defending the historical accuracy of the Old Testament. He became skeptical of science and began promoting a relatively recent origin of humans in Central Asia, as claimed by famed paleontologist Henry Fairfield Osborn. Wright's "The Passing of Evolution" (Chapter 69) was one of the few articles in *The Fundamentals: A Testimony to Truth* (1910) that attacked evolution. As he neared the end of his career, Wright advocated miracles, God, and special creation while shunning being called an "evolutionist." At the Scopes Trial, Bryan cited "a man named Wright, who taught at Oberlin" while he was being questioned by Darrow about Earth's age.

Wright died in Oberlin, Ohio, on April 20, 1921, of cardiac asthma. He is buried in Oberlin's Westwood Cemetery beneath a tombstone describing him as a "Geologist."

Sources

Numbers, Ronald L. 1988. George Frederick Wright: From Christian Darwinist to fundamentalist. *ISIS*, 79, 624–645.
Wright, G. Frederick. 1882. *Studies in Science and Religion*. Andover, MD: Draper.

Appendix

The Scopes Family

Thomas Scopes
Birth Date: 23 June 1860
Birthplace: St. Luke, Middlesex, England
Death Date: 27 May 1945
Burial location: Paducah, McCracken County, KY
Marriage Date: 15 October 1885
Father: William Scopes (1818–1866)
Mother: Mary Turner Lucas (1822–1873)

Mary Alva Brown
Birth Date: 15 July 1865
Birthplace: Fulton, KY
Death Date: 1 October 1957
Burial location: Paducah, McCracken County, KY
Marriage Date: 15 October 1885
Father: John Thompkins Brown (1829–1898)
Mother: Martha Elizabeth Calvert (1838–1920)

Children of Thomas and Mary Alva Scopes

Charles Calvert Scopes

Birth Date: 27 September 1886
Birthplace: Jackson, TN
Death Date: 8 August 1887
Burial location: Union City, Obion County, TN

Nannie Mae Scopes

Birth Date: 10 January 1888
Birthplace: Jackson, TN
Death Date: 28 December 1988
Burial location: Paducah, McCracken County, KY
Married: Walter Gilliam Sr. (1885–1965)
Marriage Date: 25 December 1907
Children: Harold Thomas Gilliam (1909–1983), Nannie Mae Gilliam Heflin (1910–2005); Nettie Marie Gilliam (1913–1934); Walter Scopes Gilliam (1916–1982); and William Rheymond Gilliam (1926–1965)

Ethel Elizabeth Scopes

Birth Date: 24 November 1889
Birthplace: Jackson, MS
Death Date: 30 July 1982 in Corpus Christi, Nueces County, TX
Burial location: Paducah, McCracken County, KY
Married: Thomas T. Clark

Marriage Date: December 25, 1913
Children: Thomas Earl Clark (1915–1989) and Mariann Clark Sale (1919–2007)

Mary Irene Scopes

Birth Date: 2 March 1892
Birthplace: Jackson, TN
Death Date: 25 June 1989
Burial location: Paducah, McCracken County, KY
Married: Edgar L. Metzer
Marriage Date: May 27, 1914
Children: Alva Louise Metzger Wahl (1917–2009), Thomas Edgar Metzger (1919–1991), Charles Linton Metzger (1923–2009), and Mae Ellen Metzger Hodges (1925–1990)

Lela V Scopes

Birth Date: 30 October 1896
Birthplace: Paducah, McCracken County, KY
Death Date: 23 January 1989
Burial location: Paducah, McCracken County, KY
Children: None

John Thomas Scopes

Birth Date: 3 August 1900
Birthplace: Paducah, McCracken County, KY
Death Date: 21 October 1970 in Shreveport, LA
Burial location: Paducah, McCracken County, KY
Married: Mildred Elizabeth Walker
Marriage Date: 3 February 1930 in Maracaibo, Zulia, Venezuela
Children: John Thomas Scopes, Jr. (1932–) and William Clement Scopes (1936–2016)

Notes

1. Winrod was the inspiration for dictator Berzelius "Buzz" Windrip in Sinclair Lewis's *It Can't Happen Here* (1935). In Lewis's book, hero Doremus Jessup claims that fascism could flourish in places where Bryan "made the world laugh itself sick by forbidding the teaching of evolution."

2. Despite Rusby's claim, Columbia never denied admission to Tennesseans to its pharmacy school.

3. The incorporators of the Bryan Memorial University Association (BMUA; October 15, 1925) included W.C. Bailey, John L. Godsey, A.P. and Wallace Haggard, Ben and James G. McKenzie, John Raulston, F.E. Robinson, F.R. Rogers, A.T. Stewart, W.D. Upshaw, and Walter White. The incorporators of William Jennings Bryan University (WJBU; July 24, 1930) included Wallace Haggard, A.M. Morgan, and F.E. Robinson, and the founders of WJBU included A.M. Morgan, Mr. and Mrs. F.E. Robinson, and Mr. and Mrs. F.R. Rogers. The founders of William Jennings Bryan University who made "a substantial contribution" to the school included Mr. and Mrs. F.E. Robinson and Mr. and Mrs. F.R. Rogers. BMUA differed from the similarly named William Jennings Bryan Memorial Association, which commissioned the first statue to honor Bryan in Washington, D.C. That statue, sculpted by Gutzon Borglum (who also sculpted Mount Rushmore) and unveiled in Washington in 1934, was moved in 1961 to Salem, Illinois (Bryan's birthplace), where it stands today.

4. Peay won 65 percent of the vote, to White's 35 percent.

5. After representing Nebraska in the House of Representatives, Bryan was never again elected to public office.

6. Bryan and Henry Clay are the only people since the passage of the 12th Amendment who have won electoral votes in three elections yet never won the presidency. Bryan's total of 493 electoral votes is the most received by anyone never elected president.

7. In April 1973, Tennessee became the first state to enact an "equal time" law allowing other theories of creation, including the Genesis account, to be taught with evolution. In April 1975, the so-called "Genesis Bill" was declared "patently unconstitutional" by the U.S. Court of Appeals for the Sixth Circuit.

8. The Northern Methodist Church was the only organization that could remove Byrd from his job; it had assigned him to the Dayton church, and it had no policy allowing him to resign.

9. The first textbook published in the United States to include "biology" in its title was Huxley, T.H., and H.N. Martin. 1876. *Course of Practical Instruction in Elementary Biology.* New York: Macmillan.

10. Hunter's first two textbooks, *Elements of Biology* (1907) and *Essentials of Biology* (1911), like textbooks from the previous century, separated botany, zoology, and physiology. *A Civic Biology* integrated these disciplines.

11. "Wreck of the Old 97," which eventually sold more than six million copies, was the biggest selling, non-holiday record in the first 70 years of recorded music.

12. In Jordan's letter written on August 7, 1925, to John Scopes, Jordan noted that he placed an ad in the *San Francisco Chronicle* and *Los Angeles Times* and, in five days, had received $1,075 for Scopes' scholarship. Jordan sent the money to Watson Davis, the treasurer of the scholarship fund. Letter provided to author by Jeanette Gilliam Travis.

13. In the days before Zip Code, letters addressed simply to "Monkey Town, Tennessee" were sent to Dayton; see Cornelius, Richard M., and Tom Davis. 2000. *Impact: The Scopes Trial, William Jennings Bryan, and Issues That Keep Revolving.* Dayton, TN: Bryan College. 69.

14. In fact, the law had already been violated, for Epperson's school included books about evolution, including Darwin's *On the Origin of Species.*

15. The nation's last such law, one enacted in Mississippi in 1926, was nullified by the Mississippi Supreme Court on December 21, 1970.

16. Charles Darwin (1809–1882) and Francis Galton (1822–1911) shared the same grandfather, physician Erasmus Darwin (1731–1802).

17. Jordan—the former president of Indiana University (1885–1891) and the founding president of Stanford University (1891–1916)—believed, "In the life of a nation, the vital differences are not matters of education, but of hereditary potentialities." He promoted nations being governed by the same laws as herds of cattle.

18. A variety of scientific advances refuted eugenics: (1) Hardy-Weinberg calculations showed that sterilizing "defectives" would never appreciably reduce the percentage of such people in the population; (2) George Shull's demonstration that increased yields of hybrid corn (i.e., "hybrid vigor") refuted the claim that racial mixing destroyed "good" racial types; and (3) identical twins raised apart after birth had different IQs, indicating that IQ was not heritable as eugenicists had claimed.

19. The Stewart brothers also helped fund the publication in 1909 of the *Scofield Reference Bible*, a Bible edited by American attorney Cyrus I. Scofield (1843–1921). This Bible introduced many people to James Ussher's date of 4004 BCE for creation. To many readers, Scofield's notes seemed part of the Bible itself. The *Scofield Reference Bible* was the bestselling Bible among fundamentalists; by the 1960s, it had sold more than 12 million copies.

20. Gruenberg's *Biology and Human Life* (1925, p. 588) claimed that if "feeble-minded" people had "no chance to produce offspring, there would be a great saving," and his *Elementary Biology* (1929, pp. 473–474) claimed that if natural selection is valid, then we should "not build hospitals for the sick, but to let them die—or survive ... letting disease kill whom it will, we shall at last have a population that is immune to all diseases." This will "bring out the best qualities of the race and ... prevent the multiplication of the unfit."

21. Young, who had been blacklisted during the McCarthy Era, protected his identity by using the pseudonym Nathan E. Douglas.

22. Bryan's brother, Nebraska governor Charles W. Bryan (1867–1945), was the Democrat Party's candidate for vice-president in 1924. The John Davis–Charles Bryan ticket was soundly defeated in the general election by Republican incumbent Calvin Coolidge and Charles Dawes.

23. The younger Malone, who was born at one minute after midnight on January 1, was the first child born in New York City in 1931. He died on January 1, 1990.

24. In 1968, the Arkansas law banning the teaching of evolution became the basis of the Supreme Court's ruling in *Epperson v. Arkansas* (393 U.S. 97) that such laws are unconstitutional.

25. The Haymarket riots occurred in the aftermath of a bombing during a labor protest in May 1886 at Haymarket Square in Chicago (i.e., the home of Darrow and several of the defense team's witnesses).

26. In Lewis's *Elmer Gantry*, the bestselling novel of 1927, the antagonist becomes a mob victim after speaking of "the 'monkey trial' in Dayton" (p. 390).

27. In response, Oklahoma anti-evolutionist Joe Miller sent the 115-pound chimp "Big Joe Viens" to Dayton for the prosecution. Although Mendi was playful and interacted with his fans, many spectators recoiled at their possible kinship with Viens, who was chained in a cage.

28. For example, in 1922 at the University of Kentucky—that is, at John Scopes' alma mater—president Frank LeRond McVey's (1869–1953) opposition helped defeat a bill banning the teaching of human evolution.

29. The textbook used in Scopes' general science class (in which Morgan was enrolled) was Lewis Elhuff's *General Science, First Course* (1916), which was introduced as "Exhibit 3" in the trial. Elhuff's book did not cover human evolution; its only mention of the word "evolution" is in its discussion of "Evolution of Machines" (p. 174).

30. *Inherit the Wind*'s record was not broken until April 4, 1998, when the musical *Rent* staged its 807th performance.

31. Scopes had been given 30 days (i.e., until August 20) to file the defense's bill of exceptions, but Raulston did not get the bill until September 14. Two days later, Raulston filed the bill of exceptions with clerk E.B. Ewing. Someone had marked out the typewritten "thirty" and "30" and replaced them with "sixty" and "60." On November 24, the Tennessee Supreme Court rejected the defense team's tardy bill.

32. The Arkansas law was overturned in 1968 by the U.S. Supreme Court in *Epperson v. Arkansas*, 17 years after Bogard's death.

33. Although Bryan claimed during the trial that humans belong in a unique group, it was Osborn's insistence that humans evolved independently of apes that actually *did* put humans in such a group.

34. Piltdown Man was "discovered" in 1912 near Piltdown, England, by amateur archaeologist Charles Dawson. Dawson announced his discovery with much fanfare as the "missing link" between apes and humans that documented ancient humans living in Europe. In 1953—that is, 41 years after its discovery—Piltdown Man was exposed as a forgery consisting of part of a human skull and a mandible from an orangutan. Dawson was the likely forger.

35. In the court transcript, this church is mistakenly listed as the Glensley Avenue Church of Christ.

36. The title page of this book lists George E. Price as the author. Thereafter, Price adopted his mother's maiden name as his middle name. In all of his subsequent writings, Price dropped "Edward" and became known as George McCready Price.

37. At the time of the Scopes Trial, Rappleyea was the post-commander of the Dayton American Legion chapter. When Bryan died five days after the trial, Rappleyea dressed in his military uniform and boots to help load Bryan's casket onto the funeral train.

38. According to Dwight Eisenhower, Higgins was "the man who won the war for us." Even a bitter Adolf Hitler praised Higgins as the "New Noah." See Brinkley, Douglas. 2000. The man who won the war for us. *American Heritage*, 51 (3), 49.

39. Riley's first published attack on evolution was his *The Finality of the Higher Criticism; or, The Theory of Evolution and False Theology* (1909). Two chapters of this book were later published as pamphlets: *The Scientific Accuracy of the Sacred Scriptures* (1920) and *Darwinism; or, Is Man a Developed Monkey* (1929).

40. At the time of Riley's death, Northwestern Bible School enrolled more than 1,100 students and was the second-largest Bible school in the United States.

41. On September 12, 1918, as he began his senior year in high school, "schoolboy" Scopes had registered for the draft. He re-registered in 1941, but was never called into military service.

42. When Reece was absent for the first days of classes, his replacement was Doris Darwin, a medical student from the University of Louisville who told reporters, "I accept evolution, but I won't teach it here. I'm no fundamentalist."

43. The court was apparently looking for a way to end the case, for Scopes' prosecutors and defense team had agreed that Raulston should impose the fine and that they would not raise the issue on appeal.

44. On March 18, 1925, Peabody's founding president Bruce Payne (1874–1937) had led a delegation demanding that Governor Peay veto the Butler Act.

45. The 37-year-old Jones died on April 30, 1900, in a train wreck in Mississippi.

46. This reception was hosted by Adele and George Shibley, who lived at 196 Bedford Lane in Dayton. For photos, locations, and information about other sites associated with the Scopes Trial, see Moore, Randy. 2018. *A Field Guide to the Scopes Trial.* Dayton, TN: Rhea County Historical and Genealogical Society.

47. Straton was a literature professor at Baylor during the junior and senior years of undergraduate student J. Frank Norris, another future and controversial leader of the antievolution movement.

48. Curtis's citation of "4004 BC" inaccurately referenced a calculation by John Lightfoot (1602–1675), a contemporary of Ussher who claimed that "Man was created by the Trinity about the third houre of the day, or nine of the clocke in the morning." Lightfoot did not mention creation of Earth, nor did he ever say this occurred on October 23; that detail was added by others. In Ussher's time, October 23—the autumnal equinox—was a traditional start of the year, and a day was often considered to start on the previous evening.

49. On September 12, 1924, WGN's listeners heard Judge John Caverly sentence Leopold and Loeb to life imprisonment after an impassioned defense by Darrow. Only the sentencing (and not the rest of the trial) was broadcast.

50. This much-repeated story may be apocryphal, for there is no verifiable evidence that it occurred.

Bibliography

Allem, Warren. 1959. *Backgrounds on the Scopes Trial at Dayton, Tennessee*. Master's Thesis, University of Tennessee, Knoxville, Tennessee.

Carpenter, Joel. 1997. *Revive Us Again: The Reawakening of American Fundamentalism*. New York: Oxford University Press.

Clark, Constance Areson. 2008. *God—or Gorilla: Images of Evolution in the Jazz Age*. Baltimore: Johns Hopkins University Press.

Conklin, Paul. 1998. *When All the Gods Trembled: Darwinism, Scopes, and American Intellectuals*. New York: Rowman & Littlefield.

Darrow, Clarence. 1932. *The Story of My Life*. New York: Scribner's.

de Camp, L. Sprague. 1968. *The Great Monkey Trial*. New York: Doubleday.

FitzGerald, Frances. 2018. *The Evangelicals: The Struggle to Reshape America*. New York: Simon & Schuster.

Gatewood, Willard B., Jr. 1966. *Preachers, Pedagogues, and Politicians: The Evolution Controversy in North Carolina*. Chapel Hill: University of North Carolina Press.

Ginger, Ray. 1958. *Six Days or Forever? Tennessee v. John Thomas Scopes*. London: Oxford University Press.

LaFollette, Marcel Chotowski. 2008. *Reframing Scopes: Journalists, Scientists, and Lost Photographs from the Trial of the Century*. Lawrence: University Press of Kansas.

Larson, Edward J. 1997. *Summer for the Gods: The Scopes Trial and America's Continuing Debate over Science and Religion*. New York: Basic Books.

Lawrence, Jerome, and Robert E. Lee. 1955. *Inherit the Wind*. New York: Bantam.

Lienesch, Michael. 2007. *In the Beginning: Fundamentalism, the Scopes Trial, and the Making of the Antievolution Movement*. Chapel Hill: University of North Carolina Press.

Livesay, Corinne. 2020. *William Jennings Bryan: A Difference Maker*. Dayton, TN: Bryan College.

Marsden, George M. 1980. *Fundamentalism and American Culture: The Shaping of Twentieth-Century Evangelicalism, 1870–1925*. New York: Oxford University Press.

Mencken, H.L. 2006. *A Religious Orgy in Tennessee*. Brooklyn, NY: Melville House.

Moore, Randy, and William McComas. 2016. *Images of America: The Scopes Monkey Trial*. Mt. Pleasant, SC: Arcadia Press.

Moran, Jeffrey P. 2012. *American Genesis: The Evolution Controversies from Scopes to Creation Science*. New York: Oxford University Press.

National Book Company. 1990 [1925, 1978]. *World's Most Famous Court Trial: Tennessee Evolution Case*. Dayton, TN: Bryan College.

Robinson, F.E., and W.S. Morgan. 1925. *Why Dayton—Of All Places?* Chattanooga: Andrews Printery.

Russell, C. Allyn. 1976. *Voices of American Fundamentalism: Seven Biographical Studies*. Philadelphia: Westminster.

Scopes, John T., and James Presley. 1967. *Center of the Storm: Memoirs of John T. Scopes*. New York: Holt, Rinehart and Winston.

Tompkins, Jerry R. 1965. (Editor). *D-Days at Dayton: Reflections on the Scopes Trial*. Baton Rouge: Louisiana State University Press.

Index

Page numbers in **bold** indicate pages with main entries.
Page numbers in ***bold italics*** indicate pages with illustrations.

AAAS *see* American Association for the Advancement of Science
ACLU *see* American Civil Liberties Union
Adam and Eve 11, 17, 193, 194
age of Earth 25, 46, 98, 129, 130, 148, 190, 195, 201*n*48; *see also* day, length of; 4004 BCE; Price, George McCready; Ussher, James
Alleged (movie) **9**, 58
American Anti-Evolution Association 11
American Association for the Advancement of Science (AAAS) 10, 56, 118
American Civil Liberties Union (ACLU) x, xii, **9–10**, 26, 27, 33, 39, 43, 57, 59, 69, 71, 77, 118, 134, 160, 163, 193
American Legion 77, 141, 145, 192, 201*n*37
American Museum of Natural History (AMNH) 117, 118, 119, 160, 185; *see also* Osborn, Henry Fairfield
AMNH *see* American Museum of Natural History
Andrews, A.B. 68, 150, 153, 189
anniversaries of Scopes Trial 49, 73, 83, 144, 166, 177–180, 192
anti-evolution organizations **10–13**, 16, 186, 187
Arkansas xiii, 11, 52–55, 97, 103, 116, 124, 132, 148, 166, 182, 200*n*24, 200*n*32; Supreme Court 54; *see also Epperson v. Arkansas*
Arlington National Cemetery (Arlington, Virginia) xii, 20, 27, 87, 137, 153

Bailey, W.C. 15, 105, 182, 199*n*3
Baldwin, Roger Nash 9, 10; *see also* American Civil Liberties Union
The Baltimore Sun 101, 102, 104
bankruptcy 51, 93, 94
Baptists 12, 30, 38, 49, 59, 80, 82, 89, 90, 95, 97, 114, 115, 116, 123, 125, 130, 132, 146, 147, 161, 172, 186, 186, 187
Benson, James 63, 64
Bible Crusaders of America 11
Biology and Human Welfare 18, 57
Birkhead, Agnes 124
Birkhead, L.M. 124
Bogard, Benjamin 11, 116, 132, 200*n*32
Borglum, Gutzon 155, 156, 199*n*3
Brewer, Thomas Jefferson ("T.J.") **13–*14***, 94, 180
Brown, Mary Alva 119–121, ***157***, ***173–177***, 197–198
Bryan, Grace Dexter 18, 19, 20, 29
Bryan, Mary Baird 1, 12, 15, 16, ***18–21***, 26, 27, 29, 44, 80, 145, 146, 155
Bryan, Ruth Baird 18, 70, 93
Bryan, William Jennings ix, x, xii, 4, 6, 9, 10, 11, 14, 15, ***17–20***, 21–29, ***31***, 33, 37, 40, 41, ***45***, 52, 53, 61, 63, ***67***, 68, 70, 75- 79, 80, 87, 90, 100, 103, 109, 115, 118, 120–122, 128, 136, 143, 145, 147, 150, 151, 155, 156, 158, 175, 177, 188, 189, 199*n*3
Bryan, William Jennings, Jr. 11, 15, 18, 20, **29–30**
Bryan College (Dayton, Tennessee) xiii, 11, ***14–18***, 28, 33, 52, 79, ***122***, 143, 144, 150, 153, 177, 180, 182, 199*n*3
Bryan Memorial University Association *see* Bryan College
Burbank, Luther 13, 60
Butler, John Washington ix, x, 4, 23, **30–*31***, 33, 68, 92, 180; *see also* Butler Act
Butler Act ix, xi, xii, 5, 6, 9–11, 14, 30–34, 37, 43, 50, 53, 59, 62–64, 68, 73, 92, 97, 106, 108, 110, 112, 113, 121, 132, 134, 138, 139, 141, 143, 150, 151, 160–166, 168, 184, 188, 192, 193, 201*n*44; repeal of 33–34, 73, 113, 120, 166, 168; *see also* Butler, John Washington
Buttram Cemetery (Dayton, Tennessee) 63, 66, 80, 98, 99, 109, 151, 153, 189, 194
Byrd, Howard Gale 1, ***34–36***, 124, 134, 199*n*8

Callahan, Timothy 17
Cartright, Amy Francis 16
Cartright, Lemuel 16, 125
Catholics 13, 44, 72, 86, 89, 116, 124, 164, 173, 181
Center of the Storm xiii, 127, 167
Chattanooga, Tennessee, newspapers 1, 52, 59, 64, 82, 102, 127, 134, 150, 191, 193
Chicago Tribune 42, 44, 154, 191
A Civic Biology ix, 9, **36–39**, 57–58, 74, 150, 151, 194, 199*n*10; *see also* Hunter, George William
Clarke, Edward Young 12, 13, 16, 186, 187
coal 15, 52, 136, 158; *see also* Cumberland Coal and Iron Company; Dayton Coal and Iron Company
Colby, Bainbridge **39–40**, 112
Cole, Fay-Cooper 59, ***60***, 86
Committee on Evolution 56, 118; *see also* American Association for the Advancement of Science
The Commoner 18, 22, 157
Conklin, Edwin Grant 56, 57, 58, 59, 92, 118, 131
Coolidge, Calvin 10, 26, 200*n*22
"Cross of Gold" speech (Bryan) ix, 21, 22, 41, 95
Cumberland Coal and Iron Company (Dayton, Tennessee) 13, 51, 66, 94, ***111***, 134, ***135***, 136
Curtis, Winterton 59, ***60***, 139, 190, 201*n*48

Dalhart, Vernon **40–41**, 199*n*11
Danville, Illinois 158, 171, 175
Darrow, Clarence Seward x, xiii, 1, ***4***–6, 9, 10, 14, 20, 24, 25, 29, 38, 40, ***41–47***, 48, 49, 57, 58, 62, 63, 67–74, 77, 79, 82, 85, ***91***, 95, 98, 102, 105, 106, 108, 109, 112–115, 118, 120, 125–129, 133, 136–139, 142, 143, 150–154, 160–166, 175, 177, 182, 184, 188, 190, 191, 195, 200*n*25, 201*n*49
Darrow, Ruby 44, 108
Darwin, Charles ix, 3, 15, 15, 38, 47, 56, 63, 70, 74, 100, 102, 117, 123, 130, 131, 159, 173, 188, 195, 199*n*14, 200*n*16
Darwin, Doris 140, 201*n*42
Darwin, Leonard **47–48**

Davenport, Charles 56, 58, 112, 118
Davis, John 10, 200*n*22
Davis, Watson **48–49**, 60, 177, 199*n*12; *see also* Science Service
day, length of 25, 42, 70, 129, 143; *see also* age of Earth
Dayton, Tennessee 4, 5–6, 10, 40, *52*; *see also* Buttram Cemetery; Dayton Coal and Iron Company; First Methodist Episcopal Church (North); First United Methodist Church (South); Hotel Aqua; Mansion; Rhea Central High School; Rhea County Courthouse; Rhea County Drive-In Theater; Robinson's Drug Store
Dayton Coal and Iron Company (Dayton, Tennessee) ix, x, 13, ***50–52***, 94, 95, ***111***, 134, 189; *see also* Cumberland Coal and Iron Company
Dayton Progressive Club 14, 67
Debs, Eugene 41, 175
Defenders of the Christian Faith 12
Democratic National Convention ix, 23, 41, 87, 93
Dershowitz, Alan 26, 42, 45
The Descent of Man ix, 16, 70
DeWitt Clinton High School (New York, New York) ix, 74, 84
"duel to the death" speech (Bryan) x, 4, 25

The Earth Speaks to Bryan 118, 119
Edison, Thomas 16, 40
Ellington, Buford xiii, 34
Elmer Gantry 103, 200*n*26; *see also* Lewis, Sinclair
Epperson, Susan ***52–56***, 132, 166, 199*n*14, 200*n*24
Epperson v. Arkansas xiii, 53, 54, 200*n*24, 200*n*32; *see also* Epperson, Susan
eugenics ix, 9, 37–39, 47, **56–58**, 74, 106, 118, 200*n*17, 200*n*18
Ewing, E.B. 15, 200*n*3

Falwell, Jerry 27, 116
Faubus, Orval 53, 54, 166
Ferguson, William x, 74, 141, 159, 160, 182, 193, 201*n*50
First Methodist Episcopal Church (North) (Dayton, Tennessee) 34, 35, 124; *see also* Byrd, Howard Gale
First United Methodist Church (South) (Dayton, Tennessee) **61**, 80, 97, 182; *see also* Jones, Charles R.
Five Points Methodist Church (Dayton, Tennessee) 34, 134
Florida real estate 12, 22, 80
football 63, 66, 67, 107, 133, 139–140, ***141***, 158–161
4004 BCE 190, 200*n*19, 201*n*48
Freeman, Joseph 18
fundamentalists ix, 3, 4, 9, 11–15, 23–28, 34, 36, 42, 43, 46, 52, 56, 58, 61, 52, 67–76, 86–80, 95, 96, 100–103, 115, 116, 123–124, 126, 129–134, 136, 140, 146–149, 158, 159, 167, 168, 185, 186, 188, 193, 193–195, 200*n*19, 201*n*42
The Fundamentals **61–62**, 195

The Genesis Flood 27, 130
"Genesis Law" xiii, 199*n*7
Germany 23, 57, 69, 145
Godsey, John Luke 15, **62–63**, 199*n*3
Goldsmith, William 57, ***60***
grand jury x, 3, 5, **63–64**, 99
Gruenberg, Benjamin 14, 200*n*20
Gulf Oil 164, 165

Haggard, Andrew Pleasant ("A.P.") 1, 15, 51, **64–66**, 68, 149–150, 182, 191, 199*n*3
Haggard, Wallace Callahan 15, 19, 64, ***66–68***, 71, 72, 128, 150, 153, 158, 162, 167, 168, 177, 178, 199*n*3
Haldeman-Julius, Emanuel 32, 44, **68–69**
Haldeman-Julius, Marcet 32, 44, **68–69**
Harris, Robert ("Bluch") 65
Hays, Arthur Garfield ***4***, 6, 9, 20, 25, 44, 46, 60, **69–70**, 73, 91–93, 103, 106, 109, 124, 126, 143, 184, 190
Hicks, Herbert Eugene 16, 28, 31, 66, ***67***, **71–72**, 153, 193
Hicks, Ira 72
Hicks, Sue Kerr 15, 19, 31, 33, 66, ***67***, 68, 71, **72–74**, 126, 140, 153, 167, 177, 178, 193
historical markers ***4***, 41, ***120***, 142–143, 165, 169; *see also* Scopes Trial Trail
Hitler, Adolf 12, 56, 100, 118, 201*n*38
Hoover, Herbert 28
Hotel Aqua (Dayton, Tennessee) x, 13, 91, 138, 151, 181, 182, 189
Hudson, Jack 63, 64
Hughes, Charles 10
Hunter, George William ix, 9, 36–39, 58, **74–75**, 131, 150, 194, 200*n*10; *see also A Civic Biology*

Inherit the Wind (movies) 6, 26, 73, ***77–80***, 84, 85, 93, 105, 144, 166, 177–178, 182, 183; *see also* Kramer, Stanley Earl
Inherit the Wind (play) xiii, 6, **75–77**, 81, 84, 88–89, 110, 120, 180, 181, 200*n*30

Jones, Charles R. 46. 61, **80–81**, 126, 153; *see also* First United Methodist Church (South)
Jones, Margaret Virginia ("Margo") 76, **81–82**, 88, 110
Jones, McKinley ("Mack") 140
Jordan, David Starr 13, 49, 57, 199*n*12, 200*n*17
Judd, Charles 59
jury 5, 6,29, 42, 45, 63, 64, 72, **82**, 91, 106, 116, 125, 126, 138, 142, 163, 184, 192

Kelly, Gene ***78***, 93; *see also Inherit the Wind* (movies)
Kenyon, Nellie **82–84**, 127, 177
KKK *see* Ku Klux Klan
Kramer, Stanley Earl 77–79, ***84–86***, 88–89, 110, 144, 178, 182, 185; *see also Inherit the Wind* (movies)
Ku Klux Klan (KKK) 3, 12, 16, 23, **86–88**, 89, 114, 116, 132, 185, 186; *see also* racism

Lawrence, Jerome 75, 76, 81, **88–89**, 110; *see also Inherit the Wind* (play)
Laws, Curtis Lee ix, **89**, 90
Lee, Robert E. 75, 76, 81, **88–89**, 110; *see also Inherit the Wind* (play)
Leopold-Loeb trial 6, 42, 128, 154, 201*n*49
Lewis, Sinclair 101, 103, 114, 199*n*1, 200*n*26
Lipman, Jacob 59, ***60***

Malone, Dudley Field xii, 1, ***4***, 5, 6, 10, 15, 25, 37, 46, 48, 69, 70, 72, 77, 82, **90–94**, 102, 103, 112, 118, 124, 126, 128, 143, 161, 162, 186
Mansion 13, 48, 50, 51, ***60***, ***94–95***, 109, 135; *see also* Dayton Coal and Iron Company
Maracaibo (Venezuela) 164
Marr, W.B. 23, 37
Martin, Thomas Theodore ("T.T.") 11, ***95–97***, 139, 191
Mather, Kirtley 13, 49, 59, 186, 190
Mathews, Shailer 59
McCarthy, Joseph and McCarthyism 71, 75, 88, 200*n*21; *see also Inherit the Wind* (play)
McKenzie, Benjamin Gordon ("Ben") 15, 16, 26, 44, ***67***, **97–99**, 127, 182, 183, 190
McKenzie, James Gordon 16, ***67***, 72, **99**, 127, 128, 153, 162, 182, 192, 199*n*3
McPherson, Aimee Semple **99–101**
memoirs 20, 127; *see also Center of the Storm*
Mencken, Henry Louis ("H.L.") x, 9, 49, 52, 70, 75, 83, 92, 93, 98, **101–104**, 106, 112, 154, 184, 186
Mendi, Jo ***104–105***, 200*n*27

Metcalf Maynard Mayo 6, 49, 57, 59, ***60***, **105–107**, 138
Miami, Florida 14, 15, 19, 20, 23, 24, 28, 73, 137
Miller, Arthur McQuiston **107**, 117
Milner, Lucile 9, 10
Milton, George 26
Mississippi 11, 95, 97, 124, 136, 199*n*15
modernism 3, 11, 15, 23, 24, 34–36, 49, 52, 59, 62, 89, 123, 147, 148, 149
"Monkey Town" 82, 105, 199*n*13
Morgan, John Harcourt Alexander **107–108**, 110, 111, 112, 113; *see also* University of Tennessee
Morgan, Luther Howard ("Scrappy") 6, 64, **108–109**, 182, 200*n*29
Morgan, Mamie 108
Morgan, Whitney 52, 125
Morris, Henry 27, 130
Murkett, Herbert 59
music 40, 41

The Nashville Tennessean 33, 140, 142
National Education Association 33
National Theatre (New York, New York) 76, 81, 88, **109–110**
natural selection ix, 38, 56, 57, 62, 117, 129, 132, 193, 200*n*20; *see also* Darwin, Charles
Neal, John Randolph, Jr. 6, 16, 20, 43, 49, 62, 63, 70, 108, ***110–114***, 118, 126, 128, 136, 160, 184, 185
Nelson, Wilbur 59, ***60***
Newman, Horatio Hackett 47, 57, 59, ***60***
Nietzsche, Friedrich 23, 102
Norris, John Franklyn ("Frank") 4, 23, 59, 87, **114–117**, 129, 131, 147, 201*n*47

Oak Grove Cemetery (Paducah, Kentucky) ***120***, 173, 177
On the Origin of Species ix, 16, 70, 173, 199*n*14
Orr, James 62
Osborn, Henry Fairfield 47, 56–57, 59, 91, 107, 112, **117–119**, 131, 160, 185, 195, 200*n*33

Paducah, Kentucky 24, **119–121**, 157, 158, 159, 161, 163, 164, 168, 171–173, 173–177, 197–198
Peabody College for Teachers (Nashville, Tennessee) xiii, 167, 168, ***169***, 201*n*44
Peay, Austin ix, 15, 16, 24, 32, 33, 37, 65, 99, 108, 112, ***121–123***, 193, 194, 199*n*4, 201*n*44
Porter, John William 11
Potter, Charles Francis 15, ***35***, 36, 69, 70, 97, **123–125**, 126, 186
prayer xii, 5, 16, 34, 43, 61, 80, 100, 102, 112, 124, **125–127**, 138, 153, 168
Presbyterians 16, 23, 27, 36, 72, 125, 153, 155, 159, 173, 175, 181
Presley, James Wright ("Jim") xiii, **127**, 136, 167
press passes 83, 99, **127–*128***, 192
Price, George McCready 27, **128–131**, 201*n*36
prison 42, 69, 83, 84, 136, 137
Prohibition 3, 12, 23, 191

racism 56, 57, 85, **131–133**; *see also* Ku Klux Klan
radio 3, 11, 27, 70, 71, 88, 92, 93, 100, 114, 142, 154, 155, 186, 188, 189, 191–192; *see also* WGN
Randall, Tony 76, 110
Rappleyea, George Washington ("Rapp") x, 1, 10, 13, 15, 17, 26, 354, 36, 39, 48, 51, 67, 71, 83, 94, 95, 109, ***111***, 112, 124, ***134–137***, 150, 153, 154, 160, 161, 193, 201*n*37
Rappleyea, Ova Corvin ("Precious") 94, 95, 109, 134, 137
Raulston, John Tate xii, 5, 6, 13–15, 20–29, 32, 38, 43–45, 49, 59, 61–68, 70, 72, 73, 79, 80, 82, 83, 98, 102, 106, 112, 124–126, 127–129, ***137–139***, 144, 154, 163, 177, 180, 181, 184, 190, 190–192, 199*n*3, 200*n*31
Reece, Raleigh Valentine ("Hunkey") **139–140**, 142, 158, 159, 201*n*42; *see also* football
Rhea Central High School (Dayton, Tennessee) ix, x, xi, xiii, 5, 13, 15, 16 109, 109, 139, **140–142**, 151, 158, 182, 194
Rhea County Courthouse (Dayton, Tennessee) ***xi***, 4, 5, 6, 13, 16, 45, 46, 52, 65, 97, 99, 106, 124, 141, **142–144**, 165, 179, 180, 182, 192
Rhea County Drive-In Theater (Dayton, Tennessee) 78, **144**, 177; *see also Inherit the Wind* (movies)
Rhea County Heritage and Scopes Trial Museum (Dayton, Tennessee) **see** Scopes Trial Museum
Rice, Kelso 27, 65, ***144–146***, ***162***
Riley, William Bell x, 3, 11, 12, 15, 24, 95, 115, ***146–150***, 201*n*39, 201*n*40
Robinson, Frank Earle ("F.E.") 5, 6, 14, 15, 16, 37, 46, 66, 68, 98, 104, 105, 128, **149–151**, 193, 199*n*3
Robinson's Drug Store (Dayton, Tennessee) x, ***21***, 36, 43, 51, 62, 68, 98, 105, 124, 134, ***151–153***, 160, 167, 177, ***181***, 182, 193, 194; *see also* Robinson, Frank Earle
Rodgers, J.J. 144, 177, 178, ***179***
Rogers, Frederick Richard 26, 61, ***67***, 80, 134, 145, **153–154**, 182, 189, 199*n*3
Rood, Paul 12
Roosevelt, Franklin Delano 16, 28, 30, 84, 155
Rosenwasser, Herman 59
Rusby, Henry 14, 199*n*2
Ryan, Quin Augustus **154–155**, 191, 192

Salem, Illinois 21, **155–157**, 158, 175, 199*n*3
Science League of America 13
Science Service 48, 49, 177
Scopes, Ethel Elizabeth ***157***, ***174***, 175, 177, 197–198
Scopes, Irene ***157***, ***174***, 197–198
Scopes, John Thomas ("J.T.") 6, 30, 33, 40, 45, 49, ***55***, 59, 60, 69, 82, ***83***, ***85***, ***91***, 106, ***111***, 112, 113, 120, 126, 127, 136, ***140–143***, ***157–174***, ***176***, 178, ***179***, 180, 184, 201*n*43, 197–198; *see also* football; historical markers; music; Paducah, Kentucky; Rhea Central High School; Salem, Illinois; Scopes, Irene; Scopes, John Thomas, Jr.; Scopes Mary Irene; Scopes, Mildred; Scopes, Nannie Mae; Scopes, William Clement ("Bill")
Scopes, John Thomas, Jr. 120, 164, 172, 197–198
Scopes, John Thomas, III ("Jack") 79
Scopes, Lela V ***157***, 158, 161, 163, 169, ***171–173***, ***174***, 175, 197–198
Scopes, Mary Irene ***157***, ***174***, 197–198
Scopes, Mildred xiii, ***83***, 120, 152, 164, 168, ***169***, 172, ***174***, 177, 178, 179, 180, 197–198
Scopes, Nannie Mae ***157***, ***174***, 197–198
Scopes, Thomas 119–121, ***157***, ***173–177***, 197–198
Scopes, William Clement ("Bill") 165, ***169***, 197–198
Scopes Trial Day 49, 77, 84, 141, 143, 144, 152, 166, ***177–180***, 197–198
Scopes Trial Grove (Dayton, Tennessee) **180**
Scopes Trial Museum (Dayton, Tennessee) 142, **144**, 152
Scopes Trial Play and Festival (Dayton, Tennessee) xiii, 61, 142, **180–182**, 192
Scopes Trial Trail (Dayton, Tennessee) 61, 150, 152, 153, ***181***–182, 201*n*47
Scott, Gary 33, 34
Seavey, Henry 128
Shelton, Harry J. ("Bud") 6, 37, 64, **182–183**

Shipley, Maynard 12, 186
Shreveport, Louisiana 55, 127, 136, 165, 168, 198
Shumlin, Herman 76, 81, 110
signs 38, ***96***, 99
Slaughter, Marion Try II *see* Dalhart, Vernon
Social Darwinism 23, 56; *see also* eugenics
statues 28, 46, 116, 143, 155, ***156***, 199*n*3
Stewart, Arthur Thomas ("Tom") 6, 14, 25, 59, 67, 74, 99, 106, 108, 109, 126, 128, 150, **183–185**, 190, 199*n*3
The Story of My Life (Darrow) 45, 46, 98
Straton, John Roach 10, 11, 12 13, 15, 25, 59, 116, 123, 124, 129, 131, 147, **185–187**, 201*n*47
Stribling, A.C. 125, 126, 193
suicide 51, 100, 150
Sunday, William Ashley ("Billy") 3, 25, 32, 43, 87, 100, 114, 147, **187–189**, 193
Supreme Kingdom 11, 12, 16, 186, 187

Taylor, J. William 194
Taylor, W.G. 80, 177
Tennessee Supreme Court (Nashville, Tennessee) xii, 60, 112, 113, 136, 163, 200*n*31
Theatre '55 (Dallas, Texas) 76, 81, 88; *see also* Jones, Margaret Virginia ("Margo")
theistic evolution 23, 27, 62, 195
Thomison, Walter Fairfield 26, 153, 182, **189**
Thompson, Jack R. 82, 91, 192
Thone, Frank 48, 49; *see also* Davis, Watson
Tompkins, Jerry 166, 167

United Production (United Gas) (Shreveport, Louisiana) 127, 165
United States Supreme Court (Washington, D.C.) 10, 53, 54, 58, 97, 166, 200*n*32
University of Chicago (Chicago, Illinois) xii,59, 106, 142, 164, 165
University of Illinois (Champaign, Illinois) 53, 158
University of Kentucky (Lexington, Kentucky) 71, 72, 107, 117, 120, 159, 164, 171, 200*n*28
Upshaw, William D. 15, 16, 199*n*3
Ussher, James 25, 148, **190**, 200*n*19, 201*n*48

Variety 5, 89
verdict 6, 72

Washburn, George 11, 13
Washington, D.C. 20, 26, 27, 39, 48, 49, 68, 87, 129, 145, 153, 155, 156, 165, 199*n*3
Waterhouse, Carmack 154
"We are not afraid" speech (Malone) 25, 92, 102
weather during trial 6, 83, **190–191**
WGN x, 106, 141, 142, 154, 162, **191–192**, 201*n*49; *see also* radio
Whitaker, Walter 60
White, Walter 1, 5, 14, 16, 24, 37, 67, 71, 73, 121, 122, 123, 128, 136, 141, 150, 188, **192–194**, 199*n*3, 199*n*4; *see also* Rhea Central High School
white supremacy 37, 74, 131–133
Why Dayton—of All Places? 150, 152
William Jennings Bryan University *see* Bryan College
Wilson, Woodrow 22, 39, 90
Winchester, Tennessee 15, 185
Winrod, Gerald Burton 12, 200*n*l
witnesses 3, 4, 5, 6, 10, 25, 29, 37, 44, 46, 48, 51, 57, **58–61**, 67, 69–72, 86, 91, 92, 94, 106–109, 112, 118, 121, 135, 138, 139, 150, 164, 183, 184, 186, 194, 200*n*5
women 3, 91, 147, 161
World War I 2, 3, 23, 69, 71, 88,134, 153
World War II 13, 56, 84, 88, 136
World's Christian Fundamentals Association x, 11, 15, 24, 43, 114, 115, 146–148, 185; *see also* fundamentalists; Riley, William Bell
Wright, George Frederick 62, **195**

www.ingramcontent.com/pod-product-compliance
Ingram Content Group UK Ltd.
Pitfield, Milton Keynes, MK11 3LW, UK
UKHW051930100926
8556IPUK00022B/1474